Electronic Power Control for Technicians

Electronic Power Control for Technicians

J. R. Penketh BSc, CEng, MIEE

BUTTERWORTHS TEC TECHNICIAN SERIES

First published 1982

British Library Cataloguing in Publication Data

Penketh, J.R.
Electronic power control.—(Butterworth's technician series)
1. Electric engineering
I. Title
621.3 TK145

ISBN 0-408-01154-8

Typeset by Scribe Design, Gillingham, Kent
Printed and Bound by Page Bros. Ltd., Norwich, Norfolk

Preface

This textbook covers the requirements of the Technician Education Council's levels III and IV material published in standard units U81/742, U76/361 and U76/363, and in the recommended material for inclusion in Higher Certificate programme – Electronics IV U79/622.

In addition there are many college-devised higher certificate units partly or wholly concerned with power electronics, and the book is intended to cover the requirements of the majority of those seen by the author. The text contains many worked examples and the set problems are designed to address the student to the important points made in each chapter.

The book should be of interest to students engaged on electrical and electronic programmes of study under TEC, but should also be of use to undergraduates meeting this subject for the first time.

I would like to express thanks and appreciation to the publishers for their constructive advice on the development of the book, and to Mrs Paula Davies for the excellent typing of the manuscript.

A special word of thanks also to my wife Marilyn, whose patience and encouragement have been greatly appreciated during the preparation of this book.

J.R. Penketh

Highbury College of Technology
Portsmouth

Contents

1 Thyristor devices

The subject of electronic control of electrical power grew out of the development of thermionic and gas-discharge valves, examples being the *thyratron* and the *mercury arc converter*. These devices were extensively used for such applications as variable speed drives, illumination controllers and temperature controllers. Also, these devices very largely replaced earlier systems of power control using magnetic amplifiers and tap-changing transformers. They were, however, bulky and fragile and suffered from limited life expectancy even when operated within their ratings.

In 1957, experiments on four-layer semiconductor devices at the Bell Laboratories in the USA resulted in a prototype power switching device being produced by the General Electric Company (USA). Because its electrical characteristics were very similar to those of the thyratron, this device was named the *thyristor.*

The original device acted like a unidirectional switch. When in the **off** state (i.e. non-conducting state) it resembled a practically open circuit, but when triggered into the **on** state it behaved like a rectifier diode, allowing current to pass from anode to cathode but not in the reverse direction.

Since 1957 a whole family of multi-layer semiconductor switching devices has been developed, including two, three and four terminal devices both unidirectional and bidirectional. All of these devices are classified as thyristors, and are distinguished by names such as *triac, diac, silicon controlled switch* (SCS), *silicon unilateral switch* (SUS), *gate turn-off thyristor* (GTO) and many others. The original device and its descendants are accurately described by some authors and users as the silicon controlled rectifier (SCR), but also it is more commonly but rather loosely referred to simply as a thyristor. In this book the term 'thyristor' refers to the three-terminal unidirectional SCR.

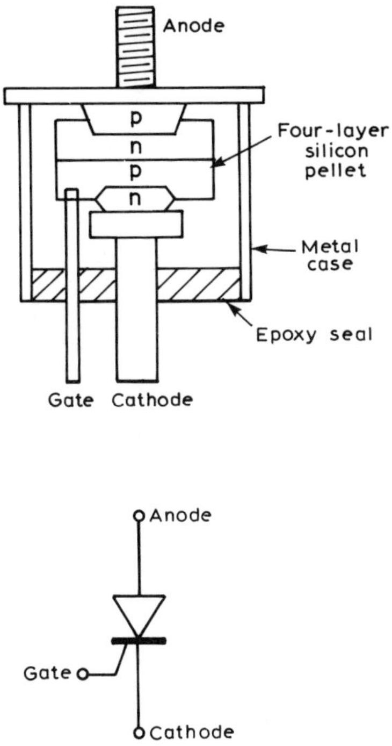

Figure 1.1 Construction (mesa type) and circuit symbol of a thyristor

THYRISTOR CHARACTERISTICS

Figure 1.1 shows the schematic construction of an alloy diffused (mesa type) thyristor and the circuit symbol.

The four-layer pnpn silicon pellet is secured firmly between two molybdenum plates, one of which forms the anode connection, often via a threaded stud for heat sink fixture, and the other forming the cathode connection. The gate connection is made to the p layer next to the cathode n layer.

One way, although not an entirely satisfactory one, of explaining the switching action of a thyristor is illustrated in *Figure 1.2.*

Figure 1.2(a) represents the four-layer device, *Figure 1.2(b)* shows the device divided into two three-layer transistors and *Figure 1.2(c)* shows the symbolic two-transistor equivalent of a thyristor.

Referring to *Figure 1.2(c)*, if A is made positive with respect to C, no current flows (except for leakage current) until G (the base of the npn transistor Tr1) is made positive. Tr1 conducts and its collector provides base current for Tr2 which in turn conducts, with its collector

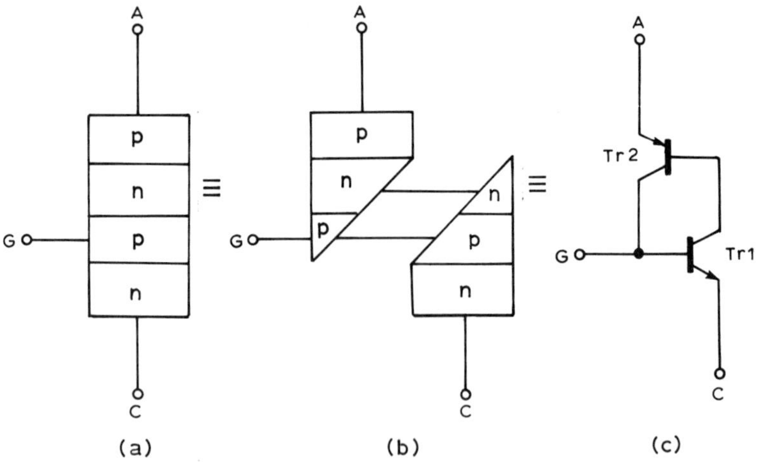

Figure 1.2 Two-transistor analogy for a thyristor

providing base current for Tr1. The original positive input to G can be removed and the conduction from A to C is self-sustaining. The principal difference between *Figure 1.2(c)* and a real thyristor is that whereas if G in this figure is made negative with respect to C, or even connected momentarily to C, conduction stops. Except for a special (and fairly new) class of thyristors, once the main anode–cathode current is flowing, the gate is unable to turn the device back into the 'off' state.

ANODE–CATHODE CHARACTERISTIC

The *static characteristic* of a thyristor is shown in *Figure 1.3.* This figure shows three distinct conditions for a thyristor. First, if a reverse voltage is applied (anode voltage V_A is negative with respect to cathode) but the gate is left open circuit a small reverse current flows (typically a few microamps). As the reverse voltage is increased a small increase in current results, but when V_A reaches a critical negative value which depends on the thyristor type there begins a very sharp rise in reverse current. This condition is known as *reverse avalanche breakdown* because it relies on the avalanche multiplication of current carriers under conditions of high

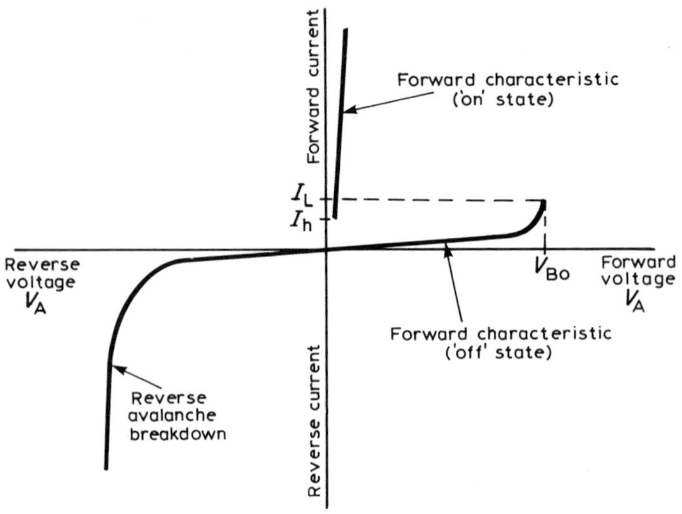

Figure 1.3 Thyristor anode–cathode characteristics

electric field in a doped semiconductor. Clearly, because the voltage across the device is large and the reverse current is large the power developed in the thyristor is considerable, and rapid destruction of the device can be expected.

Secondly, if V_A is increased in the forward direction (anode positive with respect to cathode) with no connection made to the gate, anode current increases by a few microamps in the forward direction. Near a value of V_A known as the *forward breakover voltage* (V_{Bo}) the current again starts to rise rapidly. When the anode current reaches a value called the *latching current* (I_L) the device switches on and I_L can be maintained with only a very small anode voltage (1–2 V). Clearly, in order for V_A to fall to this low value at the instant of switching there must be some form of load impedance in series with the thyristor across which the rest of the supply voltage can appear. By increasing the supply voltage further, or by altering (reducing) the load impedance, the forward current can increase in accordance with the third section of the characteristic, i.e. with only a very small increase in anode–cathode voltage. Under this condition the thyristor behaves like a near short-circuit.

If the anode current is now reduced, either by reducing the supply voltage or by increasing the load impedance, it can fall to a value called the *hold current* (I_h), below which there are insufficient current carriers to sustain the ON state and the thyristor switches to its OFF condition.

GATE CHARACTERISTICS

All of the foregoing assumed that the gate was not allowed to interfere with the anode–cathode mechanisms.

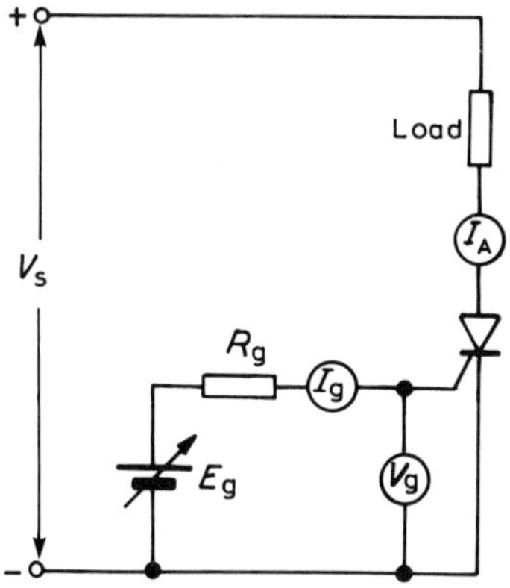

Figure 1.4 Thyristor gate test circuit

Figure 1.4 shows a simple test circuit for a thyristor, with gate drive available from a variable supply, for example via the resistor R_g. It should be noted that there are maximum and minimum limits of gate voltage and current. The maxima are fixed by the need to limit the local power developed at the p–n junction between gate and cathode (see *Figure 1.1*) and the minima are fixed by the need to achieve reliable triggering of the device.

Assuming the thyristor is in the *off state* but forward biased, as in *Figure 1.4*, an injection of sufficient current carriers by the gate will enhance the forward avalanche effect and switch the device into its *on state*. The actual internal mechanism of gate-initiated switching belongs to the realm of semiconductor physics and will not be enlarged upon here.

Taking into account the inevitable spread of diode characteristics for the gate–cathode diode and the maximum and minimum limits mentioned above, it is possible to arrive at a graphical method of determining suitable gate drive conditions for any particular thyristor type.

Figure 1.5 shows the diode characteristics of two thyristors representing the extremes of spread within one type of device (curves OM and ON). Also shown are the maximum limits of gate current and voltage and the minimum limits. Curves of maximum permissible continuous gate power and maximum permissible peak gate power are also shown. The diagram looks complicated but it is in fact fairly easy to use. The procedures are:

1. *Avoid the shaded area since this represents a region of uncertain triggering.*
2. *Ensure that the appropriate limits of current, voltage and power are not exceeded. In this connection if the gate drive is not continuous (for example, it may be a pulsed drive or a phase shifted sine wave*

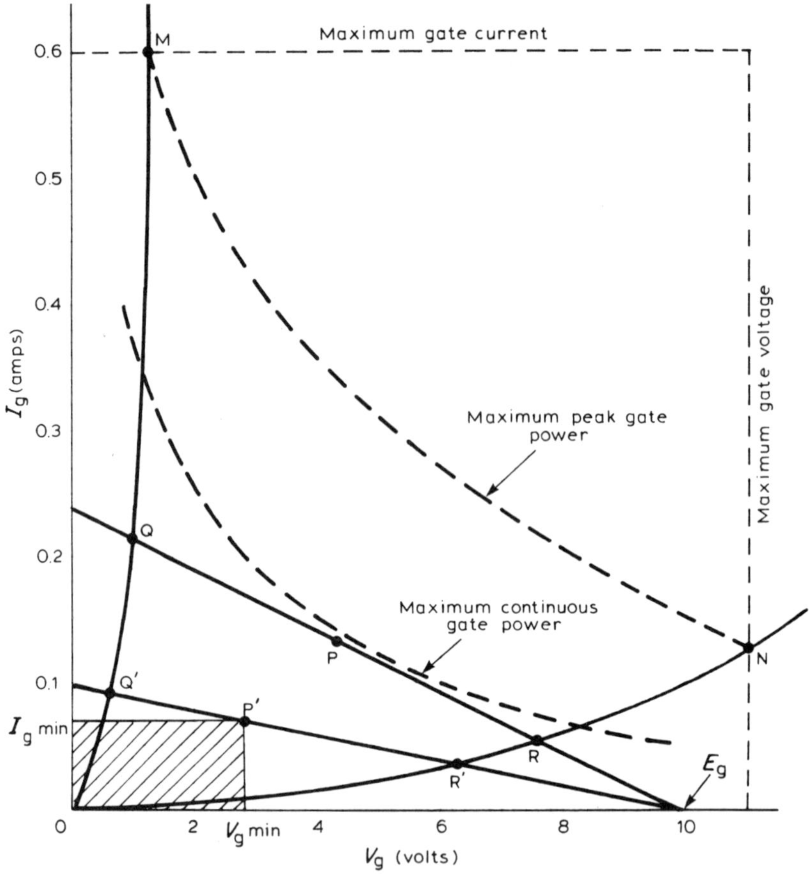

Figure 1.5 Thyristor gate characteristics

drive) ensure that neither the peak power curve is exceeded nor that the average power exceeds the continuous power curve. Ensure also the maximum values of V_g and I_g are not exceeded even for a short time.

3. *Drive the gate as hard as possible within these limits to achieve rapid switching.*

Worked example 1.1 In the circuit of *Figure 1.4* the thyristor used has the gate characteristic of *Figure 1.5*. The gate supply E_g is set at 10 V. Determine a suitable value for R_g. Also determine the maximum value of R_g.

The solution consists of finding a suitable load line. $E_g = 10$ V is one fixed point on the line and the line must come as close to the continuous dissipation curve as it can without exceeding $I_{g\,max}$ or $V_{g\,max}$. The line E_g RPQ satisfies these conditions. In fact in this case there can be no danger of $V_{g\,max}$ and $I_{g\,max}$ being exceeded.

The point P represents the gate condition for the average thyristor of this type, and points Q and R represent the two extreme cases of gate conditions.

R_g is found from the slope of the line RPQ, thus

$$R_g = \frac{10}{0.24} = 41.6\ \Omega$$

and the nearest preferred value above this is

$$\underline{R_g = 47\ \Omega}$$

The maximum value of R_g is defined by the closest approach the line E_g RPQ can make to the shaded region, that is E_g R′P′Q′ in *Figure 1.5.*

$$\therefore R_{g\,max} = \frac{10}{0.1} = \underline{100\ \Omega}$$

Worked example 1.2 A thyristor is to have a pulsed gate drive. If the curves of *Figure 1.6* are applicable, determine a suitable value for R_g. The gate voltage source is 10 V and the mark–space ratio of drive pulses is 1:3.

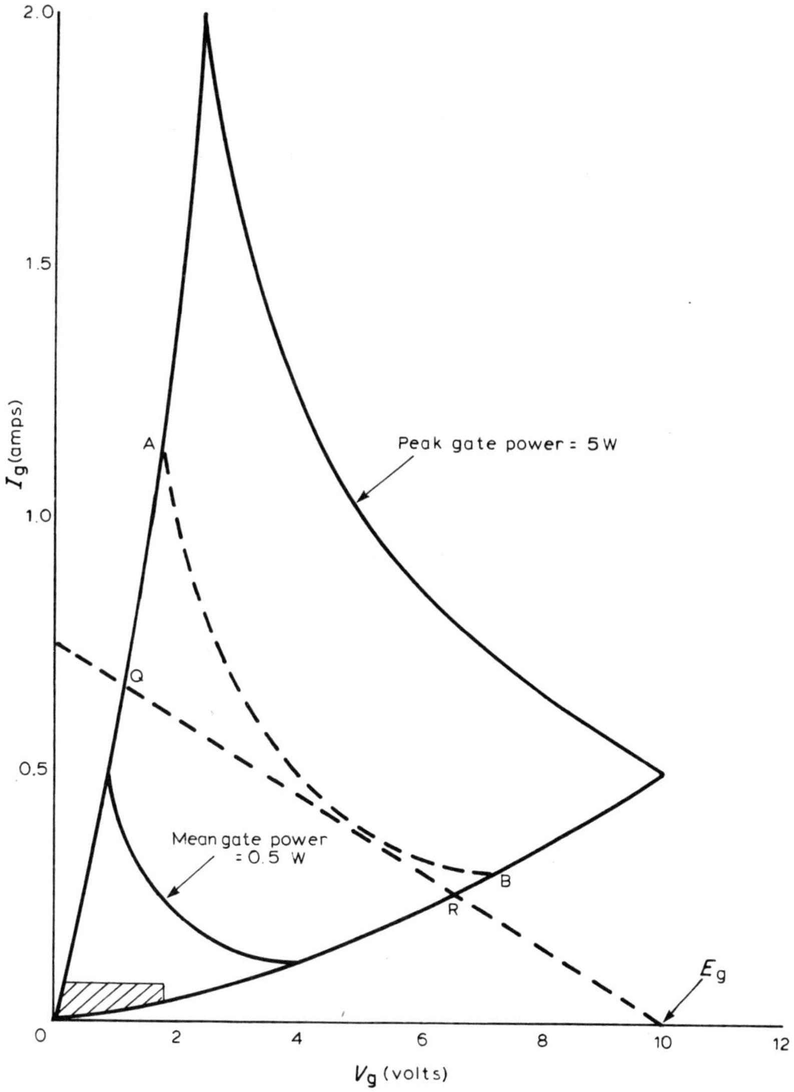

Figure 1.6 Gate characteristics for worked example 1.2 (solution shown in dashed lines

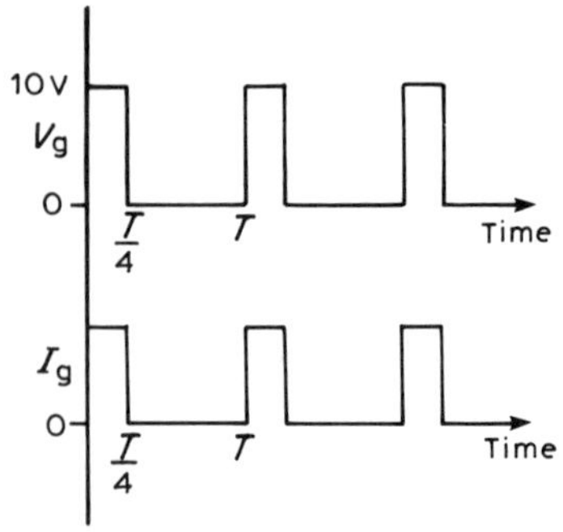

Figure 1.7 Drive waveforms for worked example 1.2

Since the gate circuit is resistive, I_g will have the same waveform as V_g. These are shown in *Figure 1.7*.

Instantaneous gate power = $V_g I_g$.

If P_p = peak gate power and P_{av} = mean gate power, then

$$P_p = V_{g\,max} \times I_{g\,max}$$

and

$$P_{av} = \frac{V_{g\,max}}{2} \times \frac{I_{g\,max}}{2}$$

since the r.m.s. values of V_g and I_g are half their peak values. Thus

$$P_p = 4P_{av}$$

The maximum value which P_{av} can take is 0.5 W from *Figure 1.6*. Thus

$$P_p = 4 \times 0.5$$
$$= 2 \text{ W}$$

The dissipation curve for 2 W is shown as dashed line AB sketched on *Figure 1.6*.

The load line drawn from E_g = 10 V must lie below line AB, and the dashed line E_g RQ is proposed.

The portion QR lies well within the current and voltage limits of 2 A and 10 V as defined by the 5 W curve.

R_g is found from the slope of RQ. Thus

$$R_g = \frac{10}{0.775}$$
$$= 12.9\ \Omega$$

The nearest preferred value above this is 15 Ω.

CHARACTERISTICS OF OTHER THYRISTOR DEVICES

As mentioned earlier, the thyristor family comprises a number of devices, some of the more common of which will be examined here.

Triac

Next to the SCR the *triac* is probably the most common thyristor. It has the ability to conduct equally well in both directions and therefore finds great application in a.c. control circuits. A schematic construction of a triac appears in *Figure 1.8*, together with the circuit symbol.

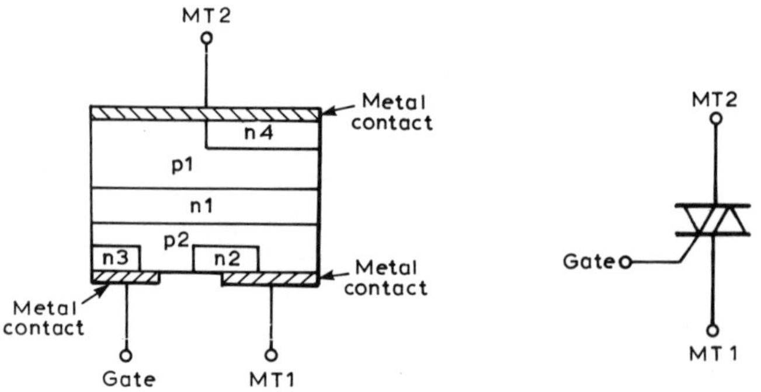

Figure 1.8 Triac construction and symbol

Since the triac is bidirectional the terms 'anode' and 'cathode' are not used, and instead these connections are labelled *main terminal 2* (MT2) and *main terminal 1* (MT1).

Static characteristics of a triac

With the gate not connected, the static characteristic resembles that of a thyristor but with a mirror image of its first quadrant replacing the third quadrant. This is shown in *Figure 1.9.*

Hold current, latching current and breakover voltage have the same significance (except for direction) as for an ordinary thyristor.

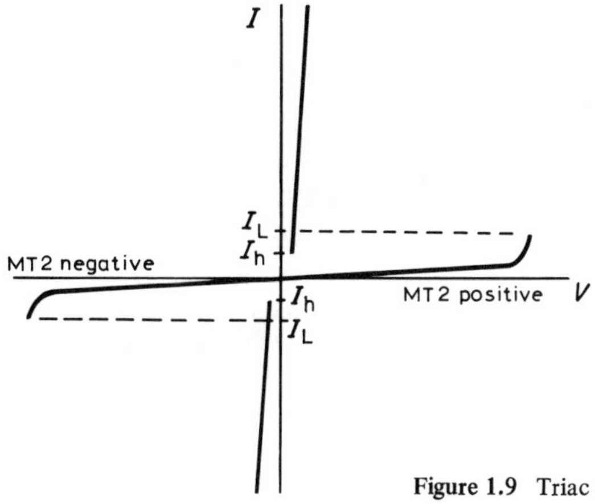

Figure 1.9 Triac static characteristic

Gate control of a triac

Consider first the case with MT2 positive. The layers n4, p1, n1, p2 form the same structure as for a normal thyristor with the gate connected to p2. A positive gate signal will turn the thyristor on as before.

If MT2 is negative (i.e. MT1 positive) the main conduction layers become n2, p2, n1, p1. Under these conditions if a negative gate signal is applied (i.e. gate made more negative than MT1) the transistor formed by n1, p2, n3 turns on (by virtue of the fact that its base, p2, is more positive than its emitter n3). This causes sufficient current carriers to appear in region n1 to break down the reverse biased junction between n1 and p1 and the device switches on. In fact, the application of either a positive or a negative gate drive in both conditions of MT2 (positive

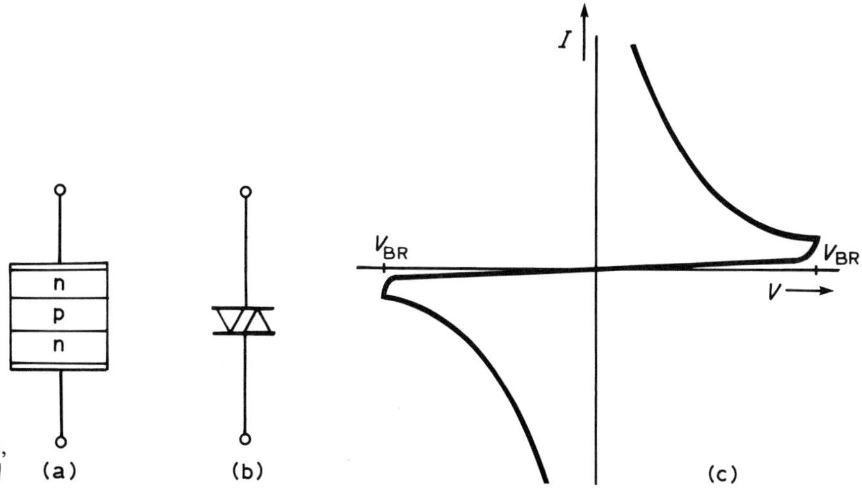

Figure 1.10 Diac construction (a), symbol (b) and characteristic (c)

or negative) will result in the device switching to the *on state*, although explanations of this based on the transistor analogies become rather tenuous.

The preferred switching modes, for minimum gate dissipation are:

MT2 positive, gate positive with respect to MT1
MT2 negative, gate negative with respect to MT1.

Diac

This is strictly not a thyristor but a transistor. However, its switching characteristics make it particularly useful in triac gate circuits so a brief description will be given here.

The construction, symbol and characteristics appear in *Figure 1.10(a)–(c)*.

The construction may be npn as in *Figure 1.10(a)*, or pnp, and some authors use an alternative symbol which identifies which type a particular device is.

The characteristic shows a well-defined breakover voltage, typically in the region 30–50 V, and a negative resistance after switching.

Light-actuated SCR (LASCR)

When silicon is irradiated with light (particularly in the infra-red range), hole-electron pairs are formed by virtue of the energy absorbed. If this occurs near the reverse biased (middle) junction of a thyristor the device can be switched into the *on* state. A recent development of this principle provides in the same encapsulation a gallium arsenide diode and a thyristor. When a small current is passed through the diode it emits light of a wavelength particularly useful for thyristor triggering by photon bombardment. This is illustrated in *Figure 1.11*.

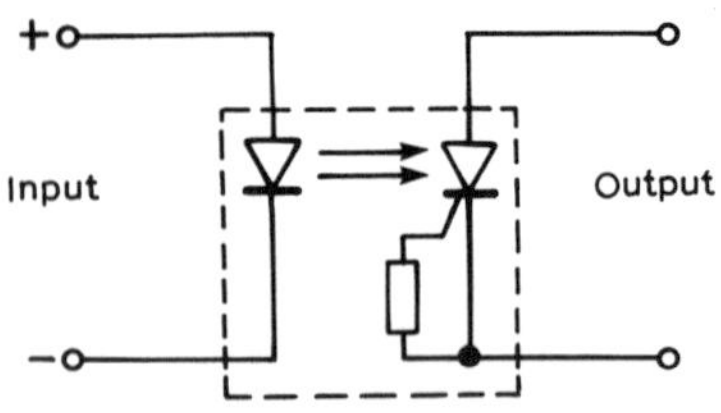

Figure 1.11 Photon coupled thyristor

These devices are variously referred to as *photon-coupled thyristors* and *opto-coupled thyristors*. They are particularly useful when isolation is required between the trigger circuit and the power circuit in a particular application. So far, only low power outputs are possible, but the opto-coupled thyristor can in turn fire a much larger device. The device is available both in SCR and triac form.

Gate turn-off thyristors

Mention is made at the beginning of this chapter of a new class of thyristors which permit the gate to control not only the turn-on but also the turn-off of the main anode current. This has been achieved mainly by careful balancing of the current gains of the two constituent transistors shown in *Figure 1.2*, resulting from advances in the various technological processes used in manufacture, among which neutron doping, passivation techniques, computer modelling and excellent lithography and process control are contributors.

Basically, referring to *Figure 1.2 (c)*, if the current gain h_{FE} of the npn transistor Tr1 can be made much larger than that of Tr2 then the same overall loop gain required for latching is achievable, but it requires much less gate current to interrupt the regenerative process than has hitherto been the case.

So far, devices having breakover voltages in excess of 1000 V and anode currents in excess of 6 A have been announced, coupled with turn-on and turn-off times of 10 μs and 1 μs, respectively, for gate drives of +1.5 V and −5 V.

Clearly, loads of several kilowatts can be controlled already by these devices, and as the technology advances a family of very high power devices particularly suitable for inverters will become available.

PROBLEMS FOR CHAPTER 1

(1) A thyristor with a d.c. supply of 200 V feeds an inductive/resistive load of time constant 10 ms and resistance 20 Ω. If the latching current of the thyristor is 20 mA determine the minimum time for which the gate drive must be applied, assuming that the internal current carrier multiplication time is negligible. *Answer*: 19 μs

(2) For a thyristor type whose gate characteristics are described by *Figure 1.6*, determine the maximum and minimum values of series gate resistance to ensure reliable triggering from a continuous gate supply of 5 V, using preferred 10% tolerance range values.

Answer: 39 Ω and 15 Ω

(3) The gate supply for the thyristor of Problem (2) is converted to a half-wave rectified sine wave of original (i.e. before rectification) r.m.s. value 8.5 V. Determine the peak gate power developed and hence a suitable value of series resistor.

Answer: Peak power = 2 W,
R_g = 22 Ω (nearest preferred above 18.2 Ω)

(4) A thyristor has an anode supply of 100 sin ωt volts, and is triggered into conduction at $\omega t = 90°$. If the load is 400 Ω and is purely resistive, and the hold current for the thyristor is 22 mA determine the angle for which the thyristor conducts during a half-cycle to the nearest degree. *Answer*: 85°

2 Power control with d.c. supplies

Many important applications of thyristors use a d.c. supply. Circuits exist using thyristors to control the mean d.c. power to a load, or to convert (or, more correctly, invert) the power available from a d.c. supply into a.c. at either a fixed or a variable frequency depending on the requirement. A typical example is illustrated in *Figure 2.1.* Here an induction motor is to be driven at varying speeds by controlling the frequency and the voltage supplied to it. The thyristor circuit inverts the d.c. supply to three-phase a.c. (usually but not necessarily non-sinusoidal waves) at a frequency dependent on the gate control circuitry. To obtain the correct voltage/frequency ratio for a particular application is fairly complicated, but the basic idea is straightforward.

Some circuits using d.c. supplies are considered in this chapter, and much of the treatment is descriptive.

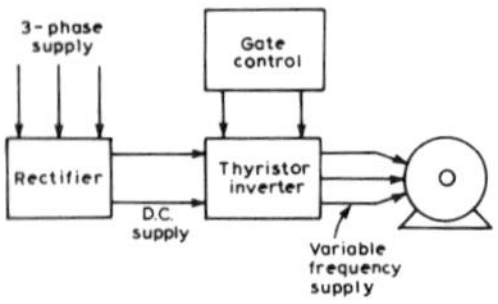

Figure 2.1 Induction motor control

SWITCHING ON AND OFF FROM A D.C. SUPPLY

A thyristor is fundamentally a *bistable device.* It is either conducting to a point of complete saturation (*the ON state*) or it is not conducting at all (*the OFF state*). In fact there is a very small leakage current when the thyristor is OFF but connected across a supply, but this has no effect on practical engineering circuits and will be ignored.

There are two methods of switching a thyristor on, one being much more desirable than the other. These methods are as follows:

1. Referring to *Figure 2.2*, if V_g is set at zero volts, so that no gate current (I_g) flows, under normal circumstances the thyristor will not conduct. However, thyristors all have a limit to which the supply voltage can be raised before the device 'breaks down'. If the supply V in *Figure 2.2* is raised sufficiently the thyristor switches into the ON state as described in Chapter 1. This method is not generally used because of the need to supply a variable d.c. supply, or to generate a high voltage pulse at the thyristor anode, and because it is not as predictable as the second method.
2. Again referring to *Figure 2.2*, if V_g is raised a gate current I_g begins to flow. When this reaches a sufficient level the thyristor conducts as described in Chapter 1. One very common way of initiating conduction is to provide a positive pulse of gate current for conduction as specified by the manufacturer.

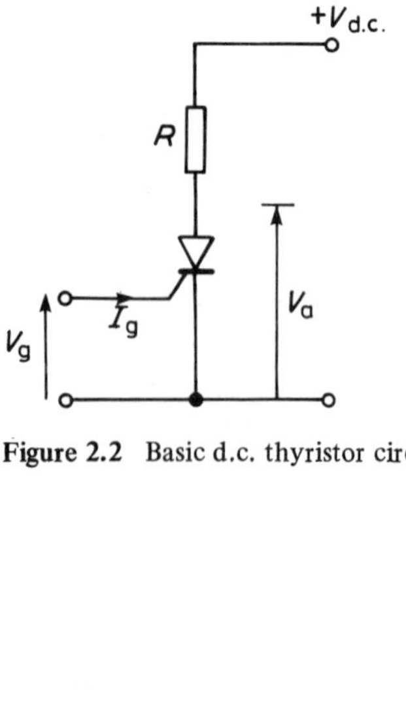

Figure 2.2 Basic d.c. thyristor circuit

Once the thyristor is conducting, the gate no longer has any control. The thyristor when ON is practically a short-circuit from anode to cathode, with typically an anode–cathode potential difference of only 1 V or less. No amount of juggling with the gate voltage and current (within the gate power limits) will alter this except for certain specialised thyristors called *gate turn-off thyristors* (GTOs). These are mentioned elsewhere in this book.

Turning a thyristor off (a process usually, but rather loosely, referred to as 'commutation') can be accomplished in two ways:

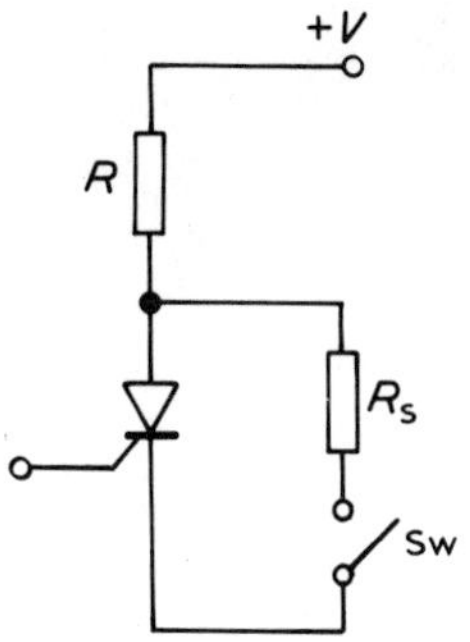

Figure 2.3 Simple turn-off circuit

1. Reduce the thyristor current to a value below the hold current (see Chapter 1). In *Figure 2.2*, if V is reduced, or if R is increased, the value of anode current will reduce. Ultimately, the thyristor will cease to conduct. Another way of achieving this kind of turn-off is shown in *Figure 2.3.* Here the switch Sw is closed momentarily and current is bled away from the thyristor and into resistor R_s. If the remaining current in the thyristor is below the hold value the thyristor turns off and remains off when Sw is opened again.
2. Remove the anode voltage or make it negative with respect to the cathode. In *Figure 2.4(a)*, if either Sw1 is opened or Sw2 is closed the anode voltage is reduced to zero. In either case conduction ceases and the device will not normally recommence conduction when the anode voltage is restored. *Figure 2.4(b)* shows an arrangement which forces the anode negative with respect to the cathode when Sw3 is closed momentarily. The thyristor ceases to conduct as before.

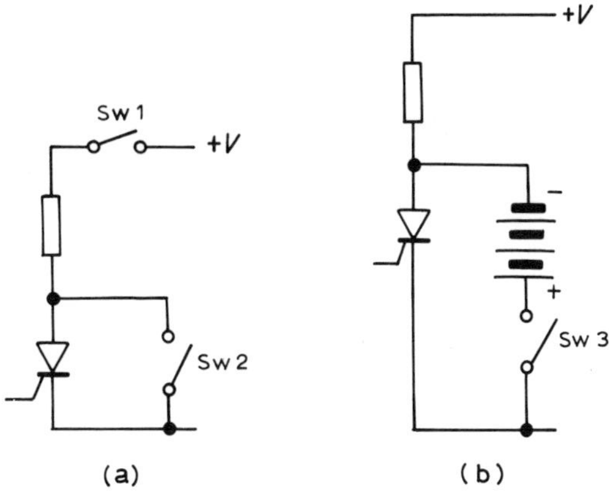

Figure 2.4 Alternative turn-off circuits

Recovery time

Just before a thyristor is switched off, a large current is passing through the device. Inside the thyristor this means that a great number of current carriers (holes and electrons) are on the move. When the device is switched off the holes and electrons still free within the device have to be given time to recombine before the anode supply is re-established, otherwise conduction will start again. Manufacturers quote a minimum commutation time (t_q) which is typically 50–100 μs, but with some thyristors is as low as 5 μs. This commutation time is quoted under certain stated conditions (for example, with a specified negative anode voltage and switching off from a specified anode current). It should be noted that recovery time is much shorter if the anode is made negative after a period of conduction rather than simply making its potential equal to that of its cathode.

Capacitor commutation (complementary commutation)

By far the most common method of commutation involves the use of a capacitor placed between the anodes of two thyristors. *Figure 2.5* shows a typical arrangement for the control of mean d.c. power in the resistor R_{L1}. The main thyristor is Th1 and the auxiliary thyristor is Th2. The circuit operates as follows.

To start with Th1 is ON and Th2 is OFF. This means that V_{A1} is practically zero and V_{A2} is at $+V$. C is fully charged to $+V$ as shown.

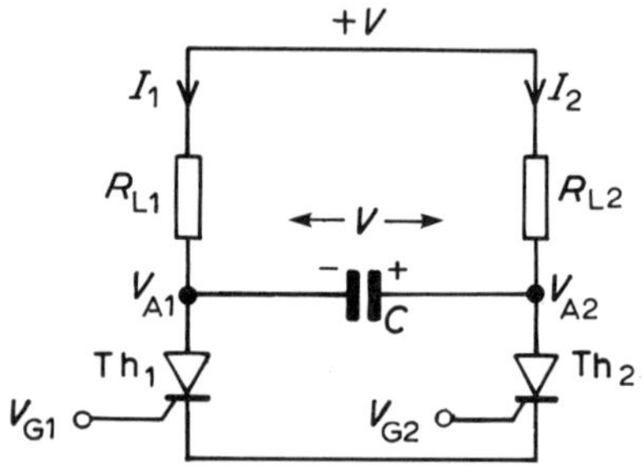

Figure 2.5 Thyristor bistable circuit

Thus

$$V_{A1} \simeq 0 \text{ V}$$
$$V_{A2} = +\text{V}$$
$$I_1 \simeq \frac{V_{A1}}{R_{L1}}$$
$$I_2 = 0$$

If now Th2 is triggered into conduction by a pulse of gate voltage and current the following events take place:

V_{A2} falls very rapidly to 0 volts
V_{A1} falls equally rapidly to $-V$ volts (because C is fully charged)
Th1 switches off because its anode is negative
V_{A1} begins to climb from $-V$ towards $+V$ on an exponential with time constant CR_{L1}.

All of this occurs extremely quickly, and indeed the operation would not succeed otherwise. The key factor is the capacitor. It behaves for a moment just like a battery, and as soon as V_{A2} drops to zero the other side of C drops by an equal amount, from zero to $-V$.

The end result of this is that Th1 is turned off and Th2 is turned on. Since the main purpose of this is to commutate Th1, Th2 and R_{L2} can be such that only a fraction of the power is drawn by these compared with Th1 and R_{L1}.

Suppose now Th1 and Th2 are triggered alternately by gate pulses V_{G1} and V_{G2}. The thyristors alternately commutate each other, with the resulting waveforms shown in *Figure 2.6.*

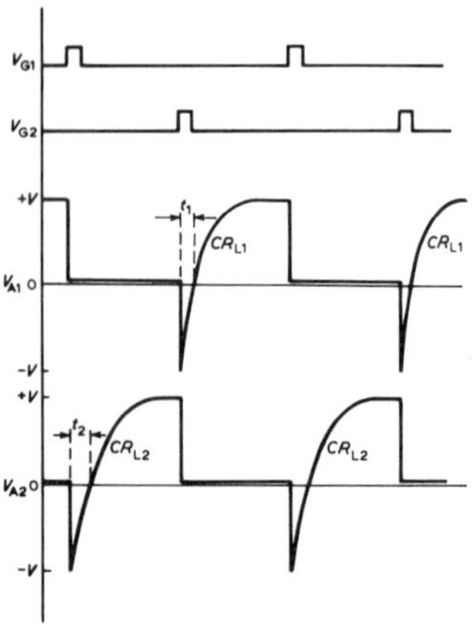

Figure 2.6 Thyristor bistable waveforms

Examining *Figure 2.6* it is clear that just after the thyristors commutate, their anode potentials rise from (practically) $-V$ to $+V$ on time constants CR_{L1} (for V_{A1}) and CR_{L2} (for V_{A2}). The time elapsed when the exponential passes zero is important, because by this time the thyristor has to be fully recovered (i.e. the hole-electron recombination complete) otherwise conduction would resume as soon as V_A went positive. Thus for each case,

$$t_1 > t_{q1}$$

and

$$t_2 > t_{q2}$$

where t_{q1} and t_{q2} are the commutation times for Th1 and Th2, respectively. Typically the recovery time allowed is made about twice the commutation time in any design, so that no problems arise because of component tolerances.

The circuit of *Figure 2.5* is often referred to as a *thyristor bistable* because either one thyristor is on or the other is on, but not both together. One further point must be noted about this circuit. At the instant when Th1 is switched off, V_{A1} is pushed down to $-V$. This means that momentarily $2 \times V$ appears across the resistor R_{L1}, and thus the instantaneous current in R_{L1} is double its normal value. All of this current flows through Th2 via C, so that the peak current in Th2 consists of the sum of this current and the current flowing through R_{L2}.

So

$$I_{2\,(\text{peak})} = \frac{2V}{R_{L1}} + \frac{V}{R_{L2}}$$

Care must be taken when selecting Th2 that its surge current rating is not exceeded.

Worked example 2.1 Design a thyristor bistable of the type shown in *Figure 2.5* using the following information:

$$\begin{aligned} V &= 100\ \text{V} \\ R_{L1} &= 10\ \Omega \\ R_{L2} &= 100\ \Omega \\ t_{q1} &= 25\ \mu\text{s} \\ t_{q2} &= 10\ \mu\text{s} \end{aligned}$$

The duty cycle is to be 50% for each thyristor.

Capacitor This is determined by the recovery time of the thyristors. Since the exponential time constants are

$$CR_{L1} \text{ for Th1}$$

and

$$CR_{L2} \text{ for Th2}$$

and since $R_{L1} < R_{L2}$, t_1 will be shorter than t_2 for any value of C.

Also Th1 requires a recovery time greater than that for Th2, so C is calculated to ensure Th1 recovers, and this automatically ensures that Th2 recovers. We have

$$t_1 = 0.69CR_{L1}$$

and making $t_1 = 2 \times t_{q1}$ for safety, this yields

$$50 \times 10^{-6} = 0.69C \cdot 10$$

$$\therefore C = 7.2\ \mu\text{F}$$

Component ratings Thyristor Th1:

$$\text{Peak current in Th1} = \frac{2V}{R_{L2}} + \frac{V}{R_{L1}}$$

$$= \frac{200}{100} + \frac{100}{10}$$

$$= \underline{12\ \text{A}}$$

$$\text{Mean current in Th1} \simeq \frac{1}{2} \times \frac{100}{10}$$

$$= \underline{5\ \text{A}}$$

(This assumes that the peak is of very short duration and then the mean d.c. current is half the ON state current for a 50% duty cycle.)

Component ratings Thyristor Th2:

$$\text{Peak current in Th2} = \frac{2V}{R_{L1}} + \frac{V}{R_{L2}}$$
$$= \frac{200}{10}\ \frac{100}{100}$$
$$= \underline{21\ \text{A}}$$

$$\text{Mean current in Th2} = \frac{1}{2} \times \frac{100}{100}$$
$$= \underline{0.5\ \text{A}}$$

Capacitor

$$\text{Peak working voltage} = \underline{100\ \text{V}}$$

$$\text{Peak repetitive current} = \frac{2V}{R_{L1}}$$
$$= \underline{20\ \text{A}}$$

Resistors

$$\text{Power in } R_{L1} = \frac{1}{2} \times \frac{V^2}{R_{L1}} \quad \text{for a 50\% duty cycle}$$
$$= \frac{1}{2} \times \frac{10^4}{10}$$
$$= \underline{500\ \text{W}}$$

$$\text{Power in } R_{L2} = \frac{1}{2} \times \frac{V^2}{R_{L2}}$$
$$= \frac{1}{2} \times \frac{10^4}{100}$$
$$= \underline{50\ \text{W}}$$

The selection of the thyristors must be made to ensure that none of the above-calculated values exceed the maximum ratings of the devices, and the rated reverse breakdown voltage must be in excess of 100 V (preferably about twice that value).

POWER CONTROL USING A BISTABLE

Referring to *Figure 2.6*, if the V_{G2} pulse waveform is moved bodily to the left or right, the time intervals for which Th1 conducts and does not conduct can be varied. If V_{G2} moves to the left, Th1 is turned off sooner and if V_{G2} moves to the right, Th1 is turned off later.

The result of this is to alter the duty cycle of Th1 (and Th2) which alters the mean power delivered to R_{L1}. There are upper and lower limits of duty cycle, however, determined by the need for each thyristor anode voltage to recover to $+V$ after it has been pushed to $-V$ by the switching action. It normally takes about five time constants for an exponential to recover fully, so V_{G1} and V_{G2} pulse separation must always allow for at least this time to elapse before switching the next thyristor.

Figure 2.7 illustrates the limits of movement of V_{G2} with respect to

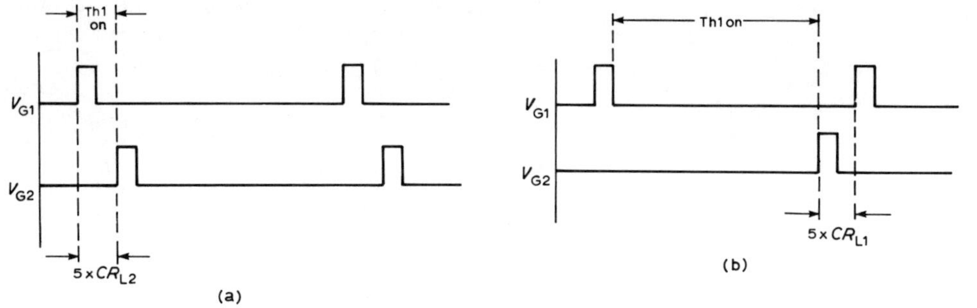

Figure 2.7 Bistable triggering limits

V_{G1}. In *Figure 2.7(a)*, V_{G2} has moved to the left, so that Th1 is turned off very soon after it is turned on, during this ON time for Th1, the Th2 anode is recovering, so the closest approach is governed by CR_{L2}.

In *Figure 2.7(b)*, V_{G2} has moved to the right, so that Th1 remains on for a long time before being turned off. In this case it is Th1 anode which has to recover in a short time, so the closeness of approach of V_{G2} to V_{G1} is determined by CR_{L1}.

Worked example 2.2 For the design in worked example 2.1, determine the maximum and minimum duty cycle for thyristor Th1, if the gate pulse repetition frequency is 100 per second.

The period corresponding to 100 pulses/s is 10 ms. The maximum ON time for Th1 is

$$T_{max} = 10 \times 10^{-3} - 5C \times R_{L1}$$
$$= 10 \times 10^{-3} - 5 \times 7.2 \times 10^{-6} \times 10$$
$$= \underline{9.64 \text{ ms}}$$

Thus the maximum duty cycle is $\frac{9.64}{10} \times 100\%$

$$= \underline{96.4\%}$$

The minimum ON time for Th1 is

$$T_{min} = 5 \times C \times R_{L2}$$
$$= 5 \times 7.2 \times 10^{-6} \times 100$$
$$= \underline{3.6 \text{ ms}}$$

Thus the minimum duty cycle is 36%.

The above example reveals a possible restriction on the choice of R_{L2}. This resistor should be as high a value as possible so as to limit the power drawn from the supply when Th1 is off. However, too high a value of R_{L2} limits the range of duty cycle available.

Rate of rise of anode voltage In Chapter 5 mention is made of the dV/dt rating of thyristors. Each thyristor type has its published maximum dV/dt rating. In the bistable circuit, the anode voltage after commutation moves exponentially past zero volts. The rate of rise at this point must not be allowed to exceed the dV/dt rating.

Worked example 2.3 In the bistable circuit of worked example 2.1, the thyristor Th1 has a dV/dt rating of 10 V/μs. Determine whether this thyristor will operate satisfactorily.

The equation governing the exponential rise of V_{A1} from $-V$ to $+V$ is

$$V_{A1} = 2V(1 - e^{-t/CR})$$

Thus

$$\frac{dV_{A1}}{dt} = \frac{2V}{CR} e^{-t/CR}$$

V_{A1} passes zero (half-way), at $t = 0.69CR_{L1}$,

$$\therefore \frac{dV_{A1}}{dt} (\text{at } t = 0.69CR_{L1}) = \frac{2V}{CR_{L1}} e^{-0.69}$$

$$= \frac{200}{7.2 \times 10^{-6} \times 10} \times 0.5$$

$$= \underline{1.381 \text{ V}/\mu\text{s}}$$

Since this result is less than the maximum dV/dt rating, the thyristor will operate satisfactorily.

BISTABLE CIRCUITS WITH INDUCTIVE LOADS

Some applications of thyristor bistables involve an inductive load in place of R_{L1}. For example, the load might be an electromagnet or the armature of a d.c. motor. This is not a particularly common arrangement, and will be dealt with only briefly here.

Figure 2.8(a) shows the basic arrangement, and *Figure 2.8(b)* separates out the inductive/capacitive equivalent circuit just after Th1 has been switched off. The switching action of the circuit is the same as with resistive loads, but the way in which the anode voltage V_{A1} recovers after switching is rather complicated.

Consider *Figure 2.8(b)*. V_{A1} started (when Th1 was ON) at zero volts, then the other side of C is suddenly taken from $+V$ to zero by the

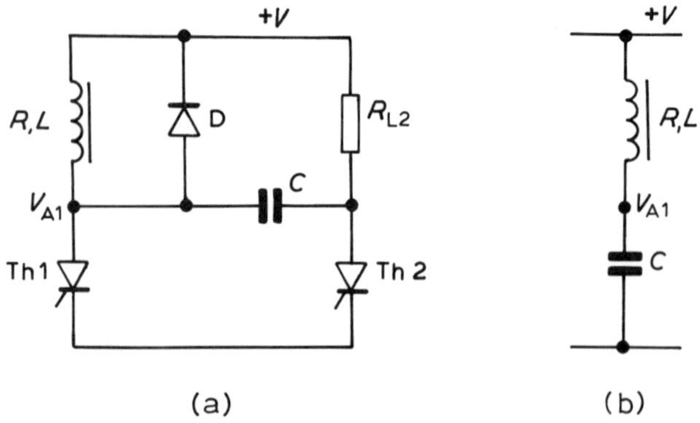

Figure 2.8 Bistable with inductive load

action of Th2 switching on. This is equivalent to applying a negative step of voltage to a series resonant circuit, with the result that V_{A1} *rings* at the damped resonant frequency. The analysis is somewhat complicated and will not be attempted here; however, the net result is that C can be made a smaller value than would be the case if the load was a pure resistor of value R.

The diode D is important. Since the *LRC* circuit is resonant, V_{A1} may well swing above the rated maximum anode voltage for Th1. However, diode D conducts as soon as V_{A1} exceeds $+V$, thereby limiting the voltage experienced by the anode of Th1.

INVERTERS

Inverters are electronic circuits which provide an a.c. power output to a load from a d.c. power supply. These constitute an extremely important class of thyristor circuits, with typical applications including the control of induction and synchronous machines and, in conjunction with a rectifier, provision of a d.c. power supply with a voltage greater than the original supply.

As with all circuits using thyristors with d.c. supplies, one critical facet of the design of an inverter will be the commutation arrangements. One common method of commutation (complementary commutation) has been discussed earlier in this chapter. There are several other methods of commutation and these will now be discussed.

Commutation of thyristors is divided into two classes. These are:

1. *Natural commutation*, sometimes called *line commutation*, in which the voltage supply is alternating. Each time the line voltage passes through zero (or, in polyphase circuits, when one phase becomes negative with respect to the next phase in sequence) a thyristor connected in the line has the opportunity to switch off. This is enlarged upon in Chapters 3 and 4.
2. *Forced commutation*, in which a specially designed network of components is incorporated to force a thyristor into its *off* state. Forced commutation methods are as follows.

Resonant commutation

In this method the load (which may be inductive, resistive, capacitive or complex) is turned into an *underdamped resonant circuit* by the addition of other components. *Figure 2.9* shows the effect of this on a simple circuit consisting of a thyristor Th and a resistive load R_L. *Figure 2.9(a)* shows the original circuit. If the thyristor is triggered *on* at $t = 0$ in *Figure 2.9(c)* the resulting load current waveform is as in graph A. It is seen to rise immediately to a value V_s/R_L, and remains at this value.

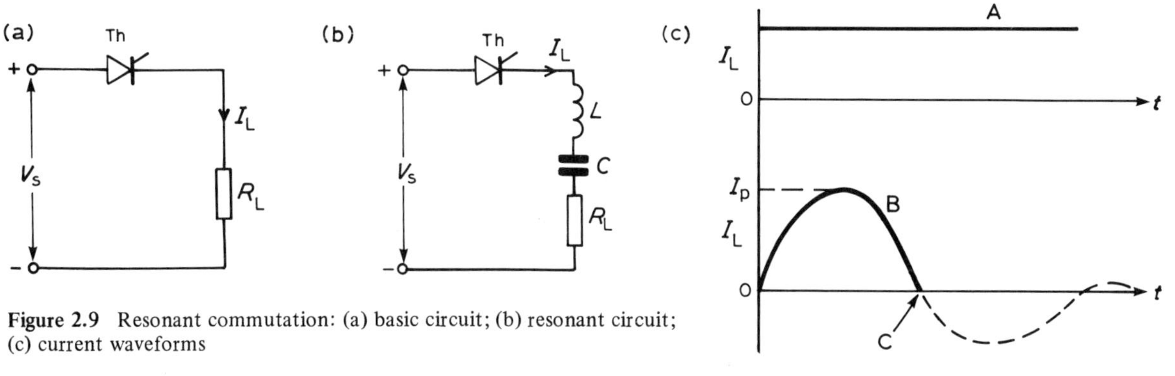

Figure 2.9 Resonant commutation: (a) basic circuit; (b) resonant circuit; (c) current waveforms

In *Figure 2.9(b)*, components L and C have been added to form an underdamped resonant circuit with R_L. When T is triggered *on*, I_L begins an oscillatory movement as in curve B. When the current reaches zero (at point C) the thyristor switches *off*.

In order to design a resonant turn-off circuit certain criteria need to be specified. These are usually the peak value of load current I_L and the time interval to commutation.

The differential equation describing the operation of the circuit after triggering, and assuming zero initial charge on C, is

$$V_s = iR + L\frac{di}{dt} + \frac{1}{C}\int i dt$$

Since this circuit is underdamped, the solution is

$$i(t) = I_p e^{-Rt/2L} \sin \omega_d t$$

where I_p is the peak value of I_L and ω_d is the damped natural frequency.

By the usual methods we have

$$I_p = \frac{V_s}{\omega_d L}$$

and

$$\omega_d = \left(\frac{1}{LC} - \frac{R^2}{4L^2}\right)^{1/2}$$

from which the *on* time interval $0C$ is found from

$$0C = \frac{\pi}{\omega_d}$$

At the end of the conduction, interval C will be fully charged to V_s, so that it will not be possible to retrigger T until the charge on C has been reduced.

Worked example 2.4 Design a resonant commutation circuit for a load resistance of 20 Ω and a supply of 100 V, giving a peak current of 4 A and an 'on' time of 5 ms.

The circuit will be as shown in *Figure 2.9(b)*. From

$$I_p = \frac{V_s}{\omega_d L}$$

we have

$$4 = \frac{100}{\omega_d L} \quad (1)$$

but

$$\frac{\pi}{\omega_d} = 5 \times 10^{-3}$$

$$\therefore \omega_d = \frac{\pi . 10^3}{5}$$

Thus, substituting in (1) and rearranging:

$$L = \frac{100.5}{4 . \pi . 10^3}$$

$$\therefore\ L = 0.04\ \text{H}$$

Also, from

$$\omega_d = \left(\frac{1}{LC} - \frac{R^2}{4L^2}\right)^{1/2}$$

we have

$$\omega_d^2 = \frac{1}{LC} - \frac{R^2}{4L^2}$$

which yields

$$C = \frac{4L}{4L^2\omega_d^2 + R^2}$$

which upon inserting the known values of R, L and ω_d yields

$$C = 5.46\ \mu\text{F}$$

Self-commutation

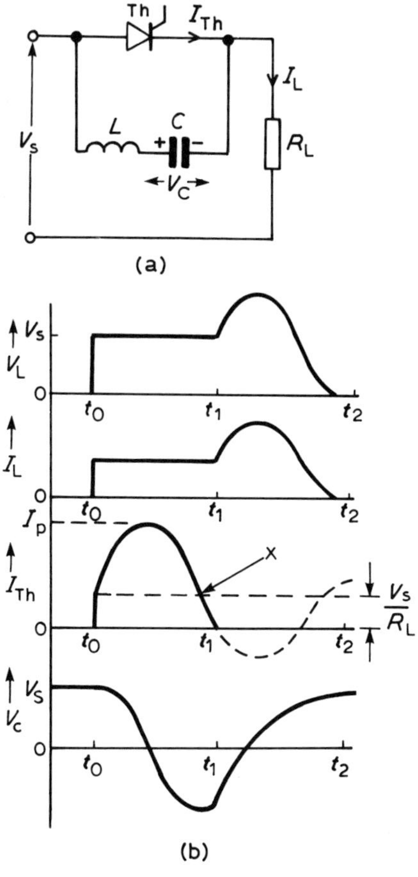

Figure 2.10 Self-commutation: (a) circuit diagram; (b) waveforms

This method of forced commutation relies again on a resonant circuit, but this time the load is not part of the resonant circuit. The basic arrangement is shown in *Figure 2.10(a)* and the voltage and current waveforms appear in *Figure 2.10(b)*.

Initially the capacitor C will be charged to V_s with the polarity shown. The thyristor Th is triggered at time t_0, and the full supply voltage appears across R_L. A resultant load current I_L flows. Since Th is a virtual short-circuit, C discharges through L and Th, forming an under-damped resonant circuit. The thyristor current therefore consists of I_L plus an oscillatory component, resulting in a high peak I_p shown in *Figure 2.10(b)*. At time t_1 the thyristor current reaches zero due to the oscillatory current, which has reversed its direction at point X, becoming equal and opposite to I_L. Th then switches off. Meanwhile the capacitor voltage V_C has reversed and has just passed its peak negative value when Th turns off.

The effect of Th turning off is to change the resonant circuit from a very lightly damped one to a more heavily damped one, since R_L now becomes the series resistance in place of Th. However, the current is still oscillatory, and at the instant t_1 it was increasing in the negative direction. This increase continues after t_1 but on a different waveform, and the current flows in R_L. The result is that both V_L and I_L increase after t_1, then fall towards zero at a rate dependent on the values of L, R_L and C. This is best understood by imagining the effect of R_L to be negligible on the damping, so that the oscillatory current continues towards a peak value (from left to right in the conductor) of twice the normal load current I_L. Thus V_L moves positively to approximately $2 \times V_s$ before falling again. Th is therefore reverse biased for approximately a quarter-cycle of oscillation, and an estimate can be made of this time compared with the necessary commutation time t_q of the thyristor (see the section on recovery time earlier in this chapter).

The *on* time from t_0 to t_1 is slightly less easy to calculate than in the resonant commutation case, since it consists of just over half of the period of oscillation of the lightly damped resonant circuit comprising

L, C and the forward resistance of the thyristor. Many practical designs, however, assume that the peak oscillatory current is twice the value of I_L (i.e. I_p is about three times the value of I_L). Under these conditions the thyristor is *on* for approximately three-quarters of cycle of the oscillatory current, which is taken to be undamped.

Worked example 2.5 Obtain the proper values of the commutating components of *Figure 2.10(a)* if the load current is 10 A, the supply is 100 V and the period of conduction T_C is 100 μs. Estimate the maximum recovery time t_q for the thyristor for successful commutation.

For an undamped oscillatory current the equation is

$$i_{(t)} = I_p \sin \omega_n t$$

where I_p is the peak value and ω_n is the undamped natural frequency $1/\sqrt{LC}$.

To evaluate I_p we equate di/dt at $t = 0$ to V_s/L, since C is initially charged to a value V_s. Thus

$$I_p \omega_n = \frac{V_s}{L} \tag{1}$$

But since

$$I_p = 2I_L$$

then

$$\frac{V_s}{\omega_n L} = \frac{2V_s}{R_L}$$

i.e.

$$\sqrt{\frac{C}{L}} = \frac{2}{R_L} \tag{2}$$

But

$$0.75 \times \frac{2\pi}{\omega_n} = T_C$$

$$\therefore \sqrt{LC} = \frac{T_C}{1.5\pi} \tag{3}$$

Multiplying (2) by (3) yields

$$C = \frac{T_C}{1.5\pi} \times \frac{2}{R_L} \tag{4}$$

Thus, inserting the value of T_C and R_L, we have

$$C = \frac{100 \times 10^{-6}}{1.5\pi} \times \frac{2}{10}$$

$$\underline{C = 4.25\ \mu\text{F}}$$

and

$$L = \frac{CR_L^2}{4} \quad \text{from (2)}$$

$$= \frac{4.25 \times 10^{-6} \times 100}{4}$$

$$\underline{L = 106\ \mu\text{F}}$$

For successful commutation the thyristor must be reverse biased for a period exceeding t_q, the recovery time.

The thyristor is reverse biased for approximately a quarter-cycle of undamped oscillation, thus

$$t_q < \frac{0.25}{0.75} \times T_C$$

$$\therefore t_q < \frac{100 \times 10^{-6}}{3}$$

$$\therefore t_q < 33\ \mu s$$

Auxiliary commutation

In the self-commutation and load resonant commutation methods, the *on* time of the thyristor is determined by the resonant characteristics of the tuned circuits, whereas the complementary method provides for *on* times dependent only on the frequency of trigger pulses applied to the gates. Another method of providing *on* times independent of the commutation components is shown in *Figure 2.11(a)*, with the associated waveforms in *Figure 2.11(b)*.

Assume initially that Th2 has been fired, resulting in C charging up to V_s with the polarity shown. Th2 turns off when this charge is complete.

At t_0 in *Figure 2.11(b)*, Th1 is fired. I_L and V_L rise immediately to the values shown, and C starts an oscillatory discharge via Th1, D and L.

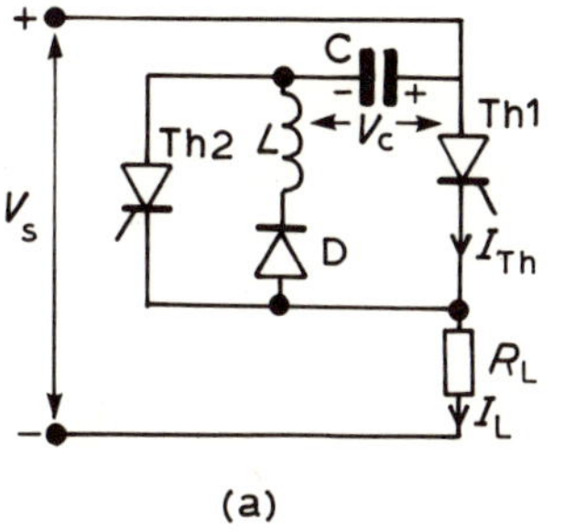

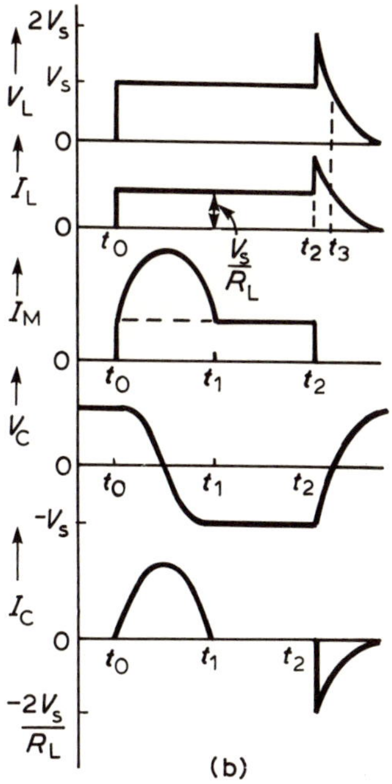

Figure 2.11 Auxiliary commutation: (a) circuit diagram; (b) waveforms

The current in Th1 is as shown. At the end of a half-cycle of oscillation at point t_1, D prevents reverse current from flowing, and C remains charged in the opposite direction. This situation endures until Th2 is fired again when the reverse voltage across C is placed directly across Th1, turning it off. Because of the stored charge on C, the voltage appearing at point t_2 across R_L is $2 \times V_s$, which reverse biases Th1 and results in current of $2 \times I_L$ flowing in R_L. Both I_L and V_L decay towards zero on time constant CR_L. The interval t_2 to t_3 represents the time for which Th1 is reverse biased, and this must exceed t_q as before.

Worked example 2.6 Design an auxiliary commutation circuit for the arrangement of *Figure 2.11(a)* in which R_L is 10 Ω, t_q for thyristor Th1 is 25 μs and the period t_0 to t_1 is to be 50 μs. The supply is 100 V.

C is determined by the time constant CR_L such that t_q is not problematical.

For safe operation, let the period t_2 to t_3 be twice the recovery time t_q:

$$\therefore 0.69CR_L = 2t_q$$

$$C = \frac{2 \times 25 \times 10^{-6}}{0.69 \times 20}$$

$$\underline{C = 7.25\ \mu F}$$

The oscillatory transient t_0 to t_1 is half of a cycle:

$$\therefore \frac{\pi}{\omega_n} = 50 \times 10^{-6}$$

$$LC = \frac{(50 \times 10^{-6})^2}{\pi^2}$$

$$L = \frac{(50 \times 10^{-6})^2}{\pi^2 \times 7.25 \times 10^{-6}}$$

$$\underline{L = 34.5\ \mu H}$$

The peak current in the diode D is

$$I_p = \frac{V_s}{\omega_n L}$$

$$= V_s\sqrt{\frac{C}{L}}$$

$$= 100\sqrt{\frac{7.25}{34.5}}$$

$$\underline{\therefore I_p = 45.8\ A}$$

The diode must have a surge rating exceeding this value, and a reverse voltage rating exceeding 100 V.

SERIES INVERTER

This type of inverter relies for its operation on the resonant turn-off circuit of *Figure 2.9.* It provides output current in the form of positive and negative pulses roughly shaped like half-sinusoids, and so long as the time separation of these is small the output current approximates to a sine wave.

The circuit operation is best understood by considering the response of a *CLR* circuit to a suddenly applied d.c. voltage. *Figure 2.12(a)* shows this arrangement. When switch Sw1 is closed, if the circuit is under-damped, the current waveform *rings*, giving rise to ringing voltages across the various circuit components. The ringing dies away on the familiar exponential with time constant $2L/R$. *Figure 2.12(b)* shows the current and capacitor voltage waveforms.

Suppose now that Sw1 is a current sensitive switch which opens when

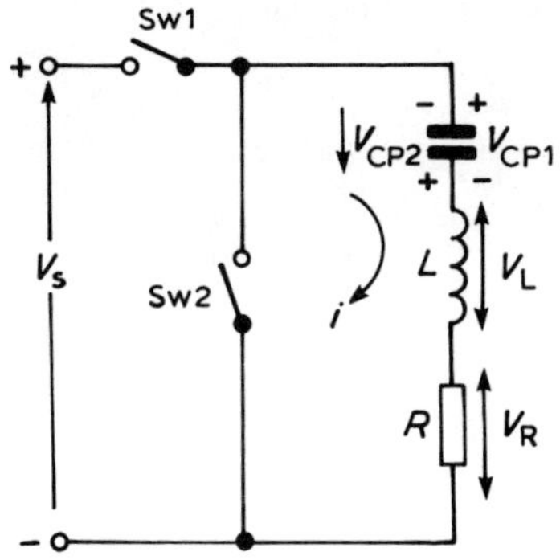

(a)

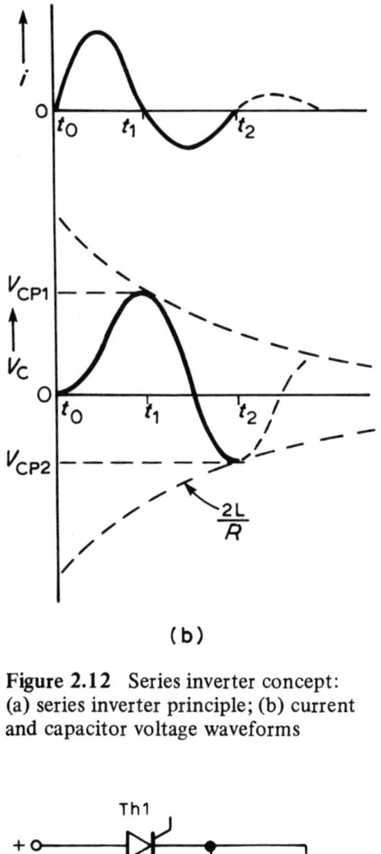

(b)

Figure 2.12 Series inverter concept: (a) series inverter principle; (b) current and capacitor voltage waveforms

the current reaches zero. In *Figure 2.12(b)*, Sw1 is closed at time t_0 and opens again at time t_1. During this time the current has gone through a positive half-cycle and the capacitor voltage V_C has risen to a value V_{CP1}. Let us 'freeze' the action at this point and take stock. The current is zero, therefore V_R is zero. V_L has reached a value in opposition to V_C. Since Sw1 has opened the current cannot continue through zero, and hence V_L collapses to zero. V_C is left 'stranded' at value V_{CP1}, with the polarity as marked in *Figure 2.12(a)*. Suppose now the second current sensitive switch Sw2 is closed. C begins to discharge via Sw2, R and L, producing another sinusoidal current pulse, this time negative. This occurs during the period t_1 to t_2 in *Figure 2.12(b)*. At the end of this time, current reaches zero, Sw2 opens, C is charged in the opposite direction to a value V_{CP2} and the voltage across L collapses to zero again. Because of the decay, $|V_{CP2}| < |V_{CP1}|$.

Now let Sw1 close again. This time V_s is connected to a system with the initial condition that

$$V_L = -V_{CP2}$$

taking into account the polarity of V_{CP2} and the direction of current flow. The total voltage now available to drive current round the circuit is initially $V_s + V_{CP2}$.

For a few cycles these conditions adjust until ultimately steady values of V_{CP2} and V_{CP1} are obtained.

Thus at the start of each positive half-cycle in the steady state the initial driving voltage is $V_s + V_{CP2}$ and at the start of each negative half-cycle the driving voltage is V_{CP1}.

Figure 2.13(a) and *2.13(b)* show the actual circuit diagram and idealised waveforms for a series inverter. Notice that there are quite definite gaps between the current pulses. This is because the thyristors,

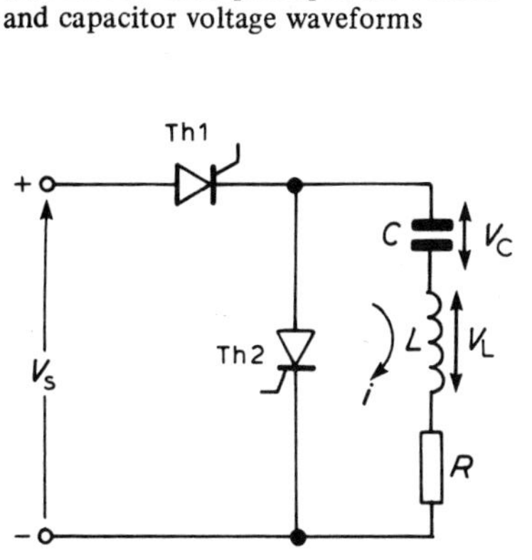

(a)

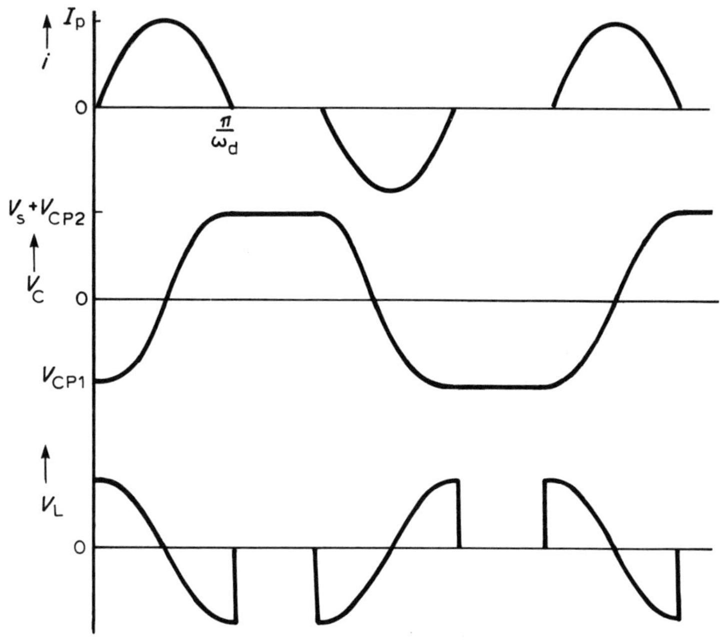

(b)

Figure 2.13 Series inverter: (a) circuit diagram; (b) waveforms

when turning off, require the usual commutation time t_q to recover before a forward bias can be applied.

Clearly, in the steady state, the charge gained by C during the positive half-cycle must be lost during the negative half-cycle, so that when things are settled

$$|V_{CP1}| = |V_{CP2}|$$

Let this steady-state condition be represented by E_C so that

$$E_C = |V_{CP1}|$$
$$= |V_{CP2}|$$

Thus during the positive half-cycle, C is charged from a value $-E_C$ to a value $V_s + E_C$.

During this time the equation governing the capacitor voltage is thus

$$V_C = -E_C + \frac{1}{C}\int_0^t i \, dt$$

The current equation will be that of a decaying sine wave, so

$$i = I_p e^{-\frac{Rt}{2L}} \sin \omega_d t$$

Thus

$$V_S + E_C = -E_C + \frac{1}{C}\int_0^{\pi/\omega_d} i dt$$

where ω_d is the damped natural frequency, i.e.

$$V_s \div 2E_C = \frac{1}{C}\int_0^{\pi/\omega_d} I_p e^{-\frac{Rt}{2L}} \sin \omega_d t \, dt$$

Evaluating this integral, after some manipulation, leads to the result

$$E_C = V_s \frac{e^{-\alpha}}{1 - e^{-\alpha}}$$

where

$$\alpha = \frac{\pi R}{2\omega_d L}$$

Consider now the reverse bias experienced by the thyristors. At the end of the first half-cycle, C is charged to $(V_s + E_C)$. Thus Th1, which has just turned off, has $+V_s$ at its anode and $V_s + E_C$ at its cathode. It is therefore reverse biased by an amount E_C. At the same time the forward bias applied to Th2 is $(V_s + E_C)$.

At the end of the negative half-cycle, Th2 has just turned off and C is charged to $-E_C$. Since V_L and V_R collapse to zero, Th2 is reverse biased by an amount E_C.

This reverse bias must be maintained long enough to clear all the hole-electron pairs before the thyristor is forward biased again.

Depending on the value of α, E_C can take any of a large range of values, and it is sometimes necessary to take advantage of this. However, a good general rule is to make E_C approximately $0.5 \times V_s$.

Thus

$$E_C \simeq V_s \frac{e^{-\alpha}}{1 - e^{-\alpha}}$$

$$\therefore \frac{e^{-\alpha}}{1 - e^{-\alpha}} \simeq 0.5$$

$$e^{-\alpha} \simeq \frac{1}{3}$$

or

$$e^{\alpha} \simeq 3$$

$$\therefore \alpha \simeq \mathbf{1.1}$$

and a satisfactory design condition is to make $\alpha = 1$.

Design procedure Using the above approach, the design procedure becomes

1. *Establish the required load condition* (R, I_P and ω_d).
2. *Determine the required value of* V_s from

$$I_P = \frac{V_s + E_C}{\omega_d L}$$

(by the usual circuit theory) and

$$\alpha = \frac{\pi R}{2\omega_d L}$$

giving

$$\omega_d L = \frac{\pi R}{2}$$

and hence

$$I_P = (V_s + E_C)\frac{2}{\pi R}$$

$$\therefore I_P = 1.5 V_s \cdot \frac{2}{\pi R}$$

and

$$V_s = \frac{I_P \pi R}{3}$$

3. *Determine the value of* L from

$$\alpha = \frac{\pi R}{2\omega_d L}$$

$$\therefore L = \frac{\pi R}{2\omega_d L}$$

4. *Determine the value of* C from

$$\omega_d^2 = \frac{1}{LC} - \frac{R^2}{4L^2}$$

hence

$$C = \frac{4L}{4L^2 \omega_d^2 + R^2}$$

5. *Determine the required ratings of the thyristors and components.*

If the inverter is to work into a variable load, the design should use the above criteria to establish the conditions for the maximum value of R. Then for the minimum value of R the new values of α and ω_d are calculated and values of peak current and reverse bias voltage on the thyristors are obtained.

In this connection it is worth noting that because output current and voltage depend intrinsically on the values of the commutating components, including R, the voltage regulation of the output (i.e. measured across R) is poor. This is a design limitation common to all series inverters.

Worked example 2.7 Design a series inverter to give a peak of 1 A into a 33 Ω resistor at a maximum frequency of 2 kHz. The thyristors have a t_q value of 50 μs.

The half-cycle occupies 0.25 ms, of which 50 μs is required for commutation. Thus the current half-sinusoid occupies 200 μs, giving a value of ω_d as

$$\omega_d = \pi \times \frac{10^6}{200}$$

$$= \underline{15.76 \times 10^3 \text{ rad/s}}$$

The required supply voltage is

$$V_s = \frac{I_P \pi R}{3}$$

$$= \frac{1 \times \pi \times 33}{3}$$

$$= \underline{34.56 \text{ V}}$$

The value of L is

$$L = \frac{\pi R}{2\omega_d}$$

$$= \frac{m \times 33}{2 \times 15.76 \times 10^3}$$

$$\underline{L = 3.3 \text{ mH}}$$

The value of C is

$$C = \frac{4L}{4L^2\omega_d^2 + R^2}$$

$$C = \frac{4 \times 3.3 \times 10^{-3}}{4 \times (3.3)^2 \times 10^{-6} \times (15.76)^2 \times 10^6 + (33)^2}$$

$$= \underline{1.1 \ \mu\text{F}}$$

The ratings are:

Thyristors Minimum reverse breakdown voltage $= E_C$
$= \frac{1}{2} V_s$

$\therefore V_{RRM} = 17.28$ V

Peak current rating = 1 A

(These conditions present no problems for any modern thyristor.)

Resistor Dissipation = 33 W.

Inductor Peak current with no saturation = 1 A.

Capacitor Current rating = 0.707 A r.m.s.

PARALLEL INVERTER

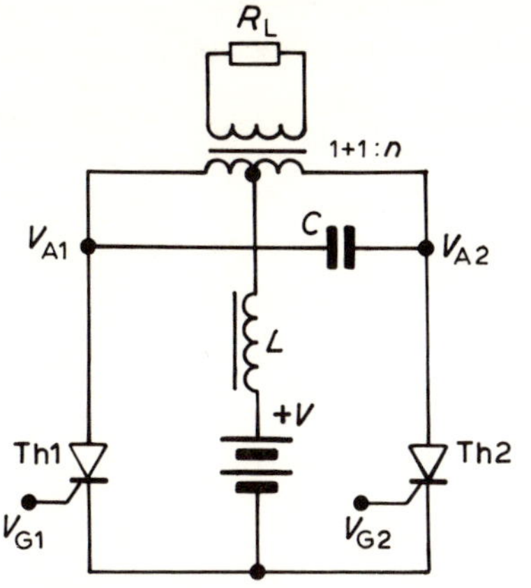

Figure 2.14 Simple parallel inverter

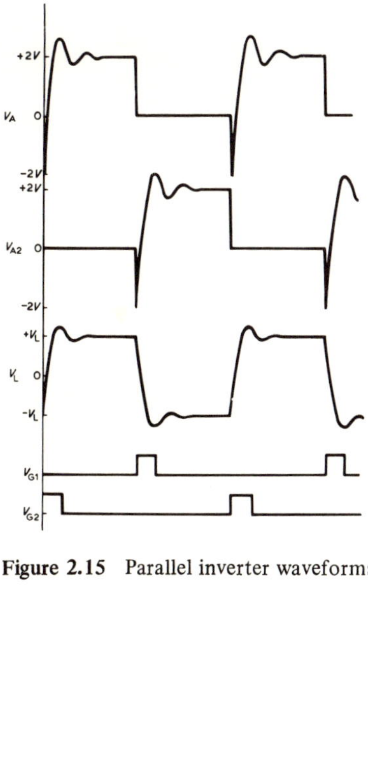

Figure 2.15 Parallel inverter waveforms

This inverter is based on the thyristor bistable. There are a number of versions, but the simplest is shown in *Figure 2.14* in which it is assumed that the gates of thyristors Th1 and Th2 are triggered alternately.

The load R_L is reflected by the transformer into the anode circuits of each thyristor, the value of the reflected load being R_L/n per anode. The purpose of the inductor L is to provide an almost constant current from the supply, rather than it being drawn in pulses. This improves the overall efficiency of the circuit. The value of L is, however, just as important as any other circuit component for correct operation.

The circuit operates as follows. Assume that Th1 is conducting, and that its anode voltage is therefore approximately at zero volts. The centre tap of the transformer will be at $+V$ volts and the anode of Th2 will be at $+2V$ volts by virtue of the transformer action. C is therefore charged to $2V$ volts. Now assume Th2 is triggered on. V_{A2} drops instantly to zero volts and V_{A1} is pushed negatively by $2V$ volts. Th1 turns off fairly quickly, and its anode voltage begins to rise again. V_{A1} exhibits a damped oscillation as shown in the wave/r.m.s. of *Figure 2.15*, eventually settling at value $+2V$.

When V_{A1} has settled, conditions are right to trigger Th1 on again, and the process repeats. The load voltage V_L is the transformed sum of V_{A1} and V_{A2}.

The operation has been analysed by Murphy and Nambiar (see References at end of chapter), who produced some simple design criteria as follows:

1. The maximum frequency of operation (for square wave) is f_{max}, where

$$f_{max} = \frac{n^2}{48CR_L}$$

This maximum frequency is obtained by the need for the ringing transient at the start of each square wave to die away (to within 5%) before the next trigger operation occurs. Frequencies lower than f_{max} may be used, but as the frequency gets lower the ability of the transformer to handle square waves becomes more critical.

2. The ratio of T_C to T_L must lie within a specified range as follows:

$$0.25 < \frac{T_C}{T_L} < 3.24$$

where

$$T_C = \frac{4CR_L}{n^2}$$

$$T_L = \frac{Ln^2}{R_L}$$

If T_C/T_L is less than 0.25, the rising edge of the square wave becomes very heavily damped so that a poor square wave results. If T_C/T_L is greater than 3.24, the ringing is so large that the anode voltage is carried negative during the first cycle, resulting in premature commutation.

3. T_C is related to the turn-off time t_q of the thyristors as follows:

$$\frac{T_C}{2} > t_q$$

If $T_C/2$ is less than t_q, the thyristor anodes are not made negative for a sufficient time to commutate the thyristors. The result would be that both thyristors remain conducting with catastrophic effects in a very short space of time.

Worked example 2.8 Design a parallel inverter (assume 100% efficient transformer) to supply 50 W at 240 V into a resistive load, using a square wave at frequency 1 kHz. A 12 V d.c. supply is available.

Assuming a perfect square wave, the r.m.s. value is equal to the amplitude.
Since the load power is 50 W, then

$$R_L = \frac{(240)^2}{50}$$
$$= \underline{1152\ \Omega}$$

Also, the turns ratio is given by

$$n = \frac{240}{12}$$
$$\therefore n = \underline{20:(1+1)}$$

Working the inverter near f_{max} will minimise transformer size, so let us assume that

$$1\ \text{kHz} = 0.8 f_{max}$$

Then

$$10^3 = 0.8\ \frac{n^2}{48CR_L}$$

i.e.

$$C = \frac{0.8 \times n^2}{48 \times R_L \times 10^3}$$

Inserting the values obtained above we get

$$C = \frac{0.8 \times 400}{48 \times 1152 \times 10^3}$$
$$= \underline{5.78\ \mu F}$$

This lies between the standard values 4.7 μF and 6.8 μF for polyester capacitors capable of handling reasonably large ripple currents.
Suppose we select a 6.8 μF capacitor.
The next stage is to determine the required t_q value for the thyristors Using condition 3 above:

$$t_q < \frac{T_C}{2}$$

i.e.

$$t_q < \frac{4CR_L}{n^2 \cdot 2}$$

$$\therefore t_q < \frac{4 \times 6.8 \times 10^{-6} \times 1152}{400 \times 2}$$

$$\therefore t_q < 39\ \mu s$$

Thyristors having t_q values of say 30 μs or less are suitable. Condition 2 provides the means for calculating the value of L, since the ratio T_C/T_L must lie between 0.25 and 3.24. The actual choice depends only on the quality of square wave required, but let us settle for a value of 3. Then

$$\frac{T_C}{T_L} = 3$$

i.e.

$$T_L = \frac{T_C}{3}$$

i.e.

$$\frac{Ln^2}{R_L} = \frac{4CR_L}{3n^2}$$

so that

$$L = \frac{4CR_L^2}{3n^4}$$

which gives on inserting values

$$L = \frac{4 \times 6.8 \times 10^{-6} \times (1152)^2}{3 \times (20)^4}$$

$$= \underline{75.2\ \mu\text{H}}$$

It should be noted that the transformer design is quite critical in that any stray inductance will modify the effective value of L. To complete the inverter a trigger circuit is needed, but this will not be dealt with here.

Load variation The simple parallel inverter described above is not very tolerant of variations of load. If R_L is altered from its design value both T_C and T_L are altered, and there is a danger that one or more of the conditions for satisfactory operation will be infringed. Because there are three inter-related conditions it is rather difficult to predict the effect of load variations and this situation becomes more complicated if the load is not purely resistive. There are, however, more advanced designs which are fairly tolerant of load variations, but these again are not within the scope of this book (see Hampson in References at end of chapter).

PROBLEMS FOR CHAPTER 2

(1) Design a thyristor bistable in which the main thyristor controls the power to a 1 kW spotlight. The d.c. supply is 100 V and the auxiliary thyristor is to draw 50 W from the supply. The thyristors have t_q values of 50 μs (main) and 30 μs (auxiliary).

Answer: $C = 14.5\ \mu$F (assuming main thyristor is reverse biased for time $2t_q$); $R_{aux} = 200\ \Omega$. Surge current for C and Th2 is 10.5 A

(2) The main thyristor of a thyristor bistable has a dV/dt rating of 5 V/μs and a t_q of 15 μs. For a load resistor of value 5 Ω and a supply of 100 V determine the minimum value of commutating capacitor so as not to violate either of these conditions.

Answer: 4.34 μF, from the t_q condition

(3) A resonant commutation circuit is required to deliver a peak of

10 A to a 5 Ω resistor from a 100 V supply, for a period of 500 μs. Determine suitable values for the inductor and capacitor.

Answer: $L = 1.6$ mH; $C = 14.7$ μF

(4) The resonant turn-off circuit of Problem (3) is modified by replacing the resistor by another of value 20 Ω. Verify that the circuit is still underdamped and determine the peak current.

Answer: Damping ratio = 0.95; $I_p = 33.7$ A

(5) Design a self-commutation circuit for a thyristor supplying a load of 33 Ω from a d.c. supply of 20 V. The thyristor is to be on for 500 μs. Determine the maximum t_q value for successful commutation.

Answer: $I = 6.4$ μF; $L = 1.74$ mH; $t_{q(max)} = 166$ μs

(6) During a test on a self-commutating circuit similar to *Figure 2.10*, it is found that when the gate is triggered the thyristor turns on and stays on. Suggest three possible reasons for this result.

(7) Design a series inverter to give 10 A peak into a 15 Ω load at a ringing frequency of $\omega_d = 10\,000$ rad/s. What is the maximum operating frequency if the thyristors require 50 μs of commutation time?

Answer: $V_s = 157$ V; $L = 2.33$ mH;
$C = 3.86$ μF; $f_{max} = 1373$ Hz

(8) Predict the resulting output conditions if the load for the inverter in Problem (7) is changed to 10 Ω.

Answer: $\alpha = 0.608$
$\omega_d = 10.99 \times 10^3$ rad/s
$E_C = 1.22$; $V_s = 191.5$ V
$I_P = 13.5$ A

(9) A parallel inverter of the type shown in *Figure 2.15* has a commutating capacitor of value 1 μF, a 1 kΩ resistive load and the turns ratio of the transformer is 12. Determine the maximum frequency of operation for square wave output, and the maximum value of t_q for the thyristors.

Answer: $f_{(max)} = 12$ kHz; $t_{q(max)} = 13.8$ μs

REFERENCES

1. Ramamourty, *An Introduction to Thyristors and their Application*, Macmillan, London, 1978.
2. Murphy, and Nambiar, 'A design basis for silicon controlled rectifier parallel inverters', *Proc. I.E.E.*, Sept. 1961.
3. Hampson, 'Application of inverter-grade thyristors in a single phase forced commutation inverter', *Mullard Technical Communications*, vol. 12, no. 115, July 1972.

3 Power control with a.c. supplies—I

In this chapter the commonly used circuits for the control of the electrical conditions in a resistive load are investigated. It is assumed that control is by triggering the thyristors into conduction at regular points within the half-cycle of supply (i.e. *phase control*). The triggering circuits may be of the *pulse generation type* or the phase *lead/lag type.* Both d.c. controllers (controlled rectifiers) and a.c. controllers are considered.

CONTROLLED RECTIFICATION

Thyristors may be used to provide a variable d.c. output from an alternating supply. This chapter deals with the common circuits for resistive loads; *Figure 3.1* shows the various basic rectifier configurations dealt with.

Polyphase versions of these are explored in less detail at the end of the chapter.

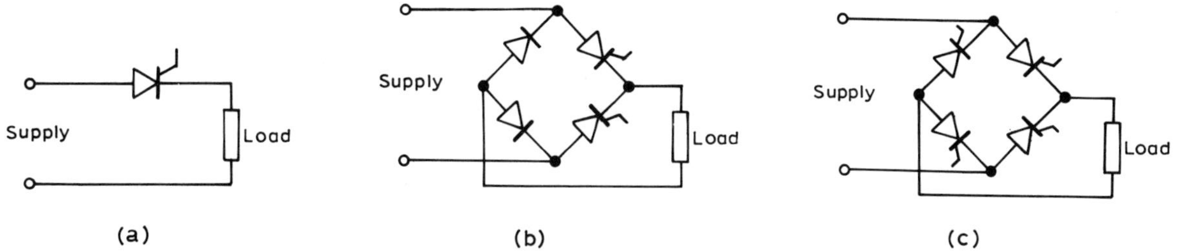

Figure 3.1 Controlled rectifier configurations: (a) half-wave rectifier; (b) half-controlled bridge; (c) fully controlled bridge

SINGLE-PHASE HALF-WAVE RECTIFIER

Figure 3.2 shows the half-wave controlled rectifier with a resistive load together with the voltage waveforms. The circuit operates as follows: V_s is applied to the circuit input terminals. The thyristor presents

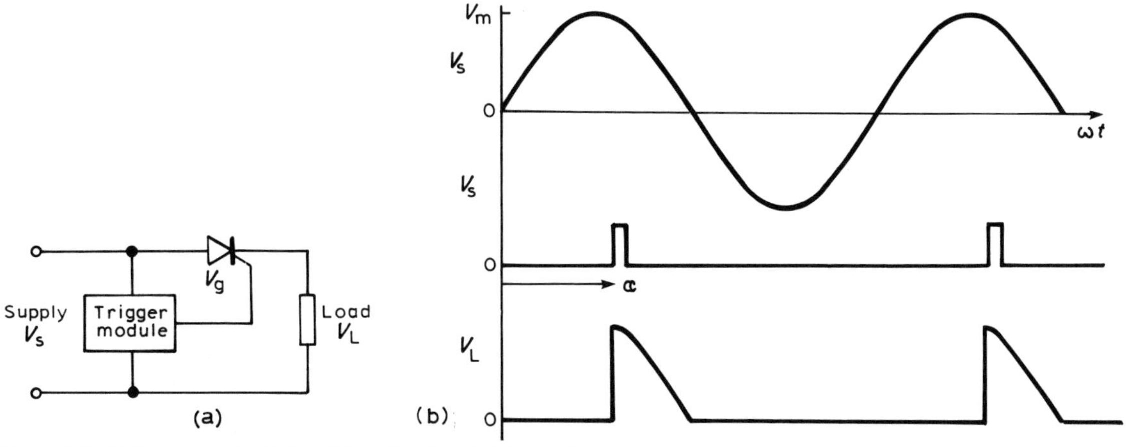

Figure 3.2 Half-wave controlled rectifier: (a) circuit; (b) waveforms

virtually an open circuit to the supply until it is triggered into conduction by the trigger module at an angle α in the positive half-cycle. When conducting, the thyristor is practically a short-circuit, except for a residual one or two volts dropped across it. Virtually the full supply voltage appears across the load after angle α. When V_s reaches zero at the end of the positive half-cycle, the thyristor turns off (commutates) and remains off until the next trigger pulse is applied during the next positive half-cycle.

More precisely the thyristor commutates when the current through it falls below the *sustaining value*, just before the end of the half-cycle, but the difference for practical purposes can be ignored.

Since the load is resistive, the load current waveform is the same shape as the load voltage waveform.

It can be seen from the load waveform that there will be an average positive d.c. value for voltage and current, and since the output is varying in time there will be an a.c. component also.

The important formulae are developed below.

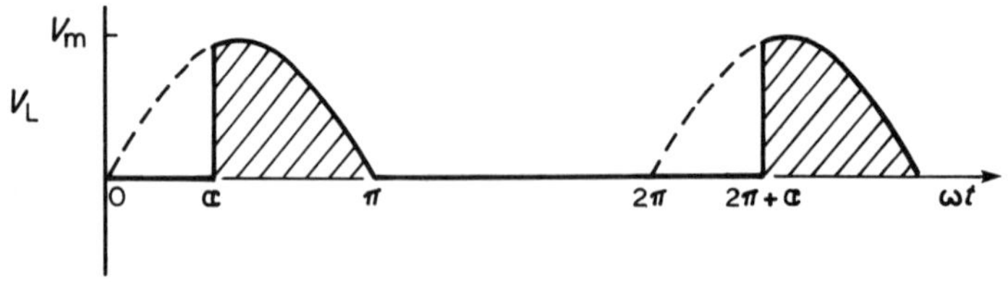

Figure 3.3 Load voltage

Average values The average load voltage is represented by the shaded area in *Figure 3.3* divided by the base length 2π, since after 2π the waveform repeats exactly. Thus

$$V_L \text{av}(\alpha) = \frac{1}{2\pi} \int_{\alpha}^{\pi} \sin \omega t . d(\omega t)$$

which yields, after inserting the limits,

$$V_L \text{av}(\alpha) = \frac{V_m}{2\pi} (1 + \cos \alpha)$$

The expression $(1 + \cos \alpha)/\pi$ occurs in several formulae and is abbreviated in this book to $g(\alpha)$. Thus

$$V_L \text{av}(\alpha) = \frac{V_m}{2} g(\alpha)$$

The average load current is

$$I_L \text{av}(\alpha) = \frac{V_L \text{av}(\alpha)}{R_L}$$

Thus

$$I_L \text{av}(\alpha) = \frac{V_m g(\alpha)}{2R_L}$$

R.M.S. values The r.m.s. load voltage is found from the standard formulae:

$$V_L \text{rms}(\alpha) = \left\{ \frac{1}{2\pi} \int_{\alpha}^{\pi} V_m^2 \sin^2 \omega t \, d(\omega t) \right\}^{1/2}$$

from which

$$V_L \text{rms}(\alpha) = \frac{V_m}{2}\left\{\frac{\pi - \alpha + \tfrac{1}{2}\sin 2\alpha}{\pi}\right\}^{1/2}$$

The expression $\{(\pi - \alpha + \tfrac{1}{2}\sin 2\alpha)/\pi\}^{1/2}$ occurs in a number of formulae and is abbreviated in the book to $f(\alpha)$. Thus

$$V_L \text{rms}(\alpha) = \frac{V_m}{2} f(\alpha)$$

and since the load is resistive:

$$I_L \text{rms}(\alpha) = \frac{V_m}{2R_L} f(\alpha)$$

Power factor The input voltage from the supply is sinusoidal, but the current flowing from the supply is not sinusoidal. There will therefore (due to harmonics generated by the arrangement) be harmonic currents flowing in the supply which do not contribute to the real power in the load. The result is a power factor less than unity even though the load is resistive.

It is important to note that the expression $\cos\theta$ which is normally regarded as representing power factor has no significance except in the special case of purely sinusoidal load voltage and current waveforms.

The expression for power factor for any circuit is

$$\text{p.f.} = \frac{\text{Active input power}}{\text{Apparent input power}}$$

If the small loss in the thyristor itself is ignored, the active input power is equal to the r.m.s. load power, and the apparent input power is the product of r.m.s. supply voltage and r.m.s. supply current. Thus

$$P_{in}(\text{app})(\alpha) = \frac{V_M}{\sqrt{2}} \times I_L \text{ rms}(\alpha)$$

since the supply current must be equal to the load current.

The r.m.s. load power is obtained from

$$P_L(\text{rms})(\alpha) = \frac{[V_L \text{rms}(\alpha)]^2}{R_L}$$

Putting these results together the power factor is

$$\text{p.f. }(\alpha) = \frac{P_L(\text{rms})(\alpha)}{P_{in}\text{app}(\alpha)}$$

$$= \frac{[V_L(\text{rms})(\alpha)]^2 \times \sqrt{2}}{R_L V_m \times I_L(\text{rms})(\alpha)}$$

$$= \frac{\left[\frac{V_m}{2} f(\alpha)\right]^2 \times \sqrt{2} \times R_L}{R_L \times V_m \times \left[\frac{V_m}{2} f(\alpha)\right]}$$

$$= \frac{f(\alpha)\sqrt{2} \times 2}{4}$$

so that

$$\text{p.f. }(\alpha) = \frac{f(\alpha)}{\sqrt{2}}$$

Design curves for resistive load circuits

The trigger angle α can be varied by the trigger module to lie anywhere between 0 and π radians. To solve the equations in order to determine the load conditions is obviously cumbersome. The problem is greatly simplified if $f(\alpha)$ and $g(\alpha)$ are plotted against α. For any value of α the appropriate function can be read off and a relatively simple calculation will then yield all the information required.

Curves of $f(\alpha)$ and $g(\alpha)$ against α are shown in *Figure 3.4*.

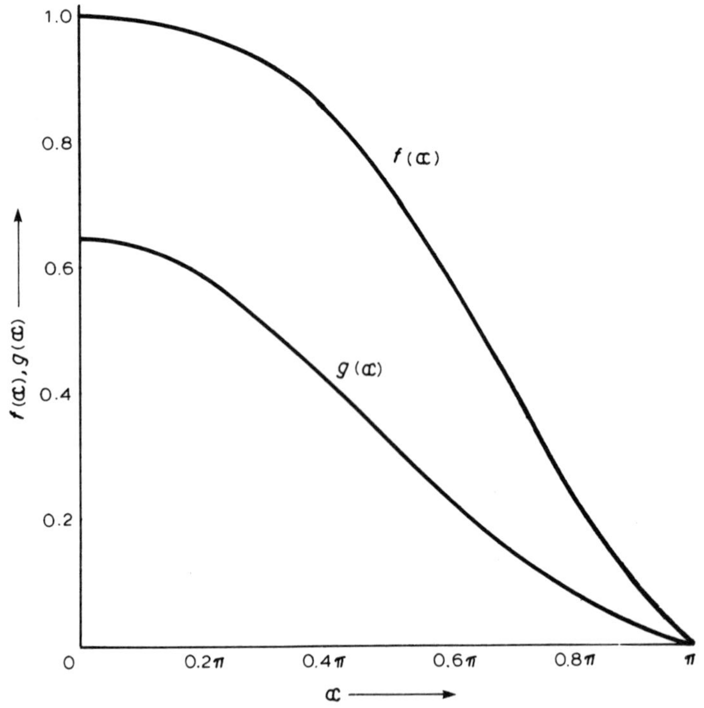

Figure 3.4 Design curves

Worked example 3.1

A small water heater having a resistance of 7 Ω which is constant with temperature is controlled by a half-wave thyristor circuit operating from a 240 V r.m.s. main supply.

For firing angles of $2\pi/3$ and $\pi/3$ determine

1. The mean d.c. load voltage.
2. The load power.
3. The power factor.

1. The mean d.c. load voltage is

$$V_L \text{av}(\alpha) = \frac{V_m}{2} g(\alpha)$$

From *Figure 3.4* for $\alpha = 2\pi/3$,

$$g(\alpha) = 0.16$$

for $\alpha = \pi/3$,

$$g(\alpha) = 0.46$$

and since $V_m = \sqrt{2} V\text{rms}$

$$V_L\text{av}\left(\frac{2\pi}{3}\right) = \sqrt{2} \times 240 \times 0.46$$
$$= \underline{54.3 \text{ V}}$$

and

$$V_L\text{av}\left(\frac{\pi}{3}\right) = \sqrt{2} \times 240 \times 0.46$$
$$= \underline{156.1 \text{ V}}$$

2. Load power is given by

$$P_L\text{rms}(\alpha) = \frac{[V_L\text{rms}(\alpha)]^2}{R_L}$$
$$= \frac{V_m^2 f^2(\alpha)}{4R_L}$$

From *Figure 3.4* for $\alpha = 2\pi/3$,

$$f(\alpha) = 0.45$$

and for $\alpha = \pi/3$,

$$f(\alpha) = 0.9$$

and using $V_m = \sqrt{2} V\text{rms}$

$$P_L\text{rms}\left(\frac{2\pi}{3}\right) = \frac{240^2 (0.45)^2}{2 \times 7}$$
$$= \underline{833 \text{ W}}$$

and

$$P_L\text{rms}\left(\frac{\pi}{3}\right) = \frac{240^2 (0.9)^2}{2 \times 7}$$
$$= \underline{3.333 \text{ kW}}$$

3. Power factor is given by

$$\text{p.f.}\ (\alpha) = \frac{f(\alpha)}{\sqrt{2}}$$

so using the $f(\alpha)$ values already obtained:

$$\underline{\text{p.f.}\left(\frac{2\pi}{3}\right) = 0.318}$$

and

$$\underline{\text{p.f.}\left(\frac{\pi}{3}\right) = 0.636}$$

Note the very considerable power factor penalty resulting from the discontinuous nature of control. Electricity generating boards are particularly interested in large-scale thyristor control of industrial plant for this reason, since the result is a good deal of harmonic current in the generating equipment and the lines for which some way has to be found to charge the customer. A further problem resulting from the generation of the higher order harmonics is the radiation from long transmission lines and resulting interference with telephone communications.

HALF-CONTROLLED BRIDGE

The two possible circuit configurations for a half-controlled bridge with resistive load are shown in *Figure 3.5* together with the voltage waveforms.

The two circuits produce identical results but the detail of the trigger module connections may differ. In circuit (a) the two thyristor anodes are electrically connected, which means the devices can be bolted to the

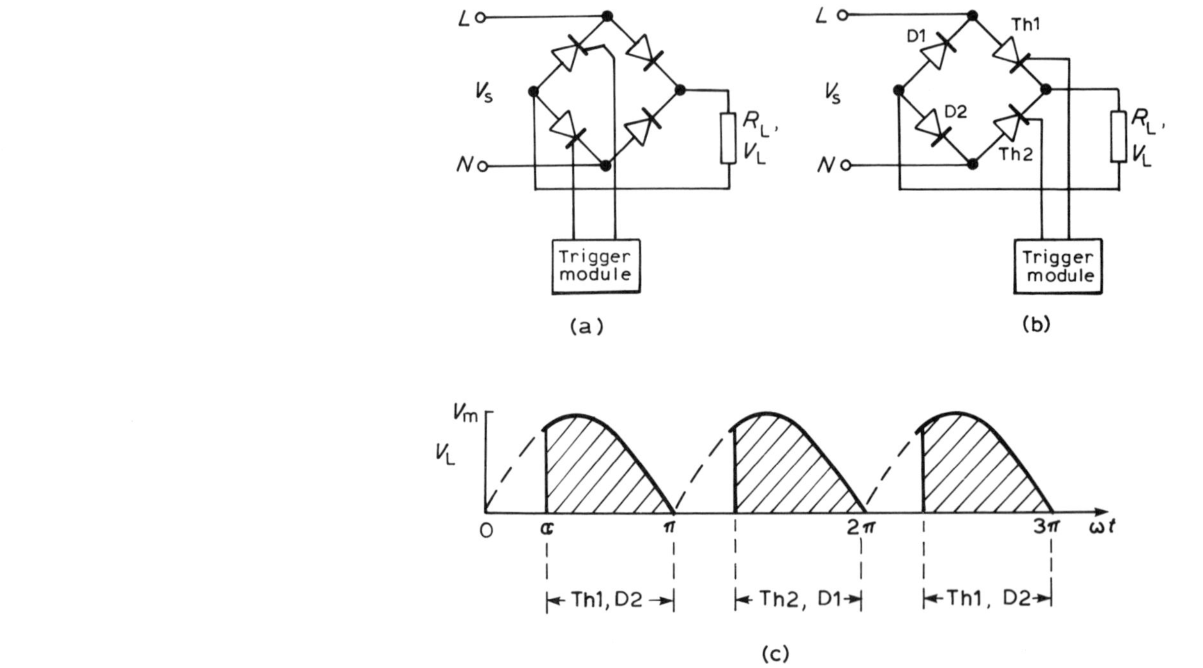

Figure 3.5 Half-controlled bridge: (a) common cathodes; (b) common anodes; (c) load voltage waveform

same heat sink without insulating washers, giving the best thermal characteristics. At the same time, since power diodes are generally constructed with their fixing nuts connected electrically to their cathodes, the diodes can also be directly mounted to their own common heat sink. This may be an advantage in low power circuits, where often common heat sinks are used.

Circuit (b), however, has the two thyristor cathodes electrically common, and this makes the trigger module design somewhat simpler. Looking at circuit (b), the operation is as follows:

When L is positive with respect to N, Th1 is triggered at an angle α. Thyristor Th1 conducts and current passes from L, through Th1, R_L and D2 back to N. When L becomes zero (i.e. the supply voltage is crossing the zero point) Th1 commutates and current ceases. L then goes into its negative half-cycle and N is effectively more positive than L. Th2 is triggered at angle α and current flows from N via Th2, R_L and D1 back to L. In both cases current flows the same way through R_L.

In *Figure 3.5* the conducting elements are marked on the waveform. Formulae for the conditions in the load are developed below.

Average values The waveform repeats every π radians, so the mean d.c. voltage is found from

$$V_L \text{av}(\alpha) = \frac{1}{\pi} \int_{\alpha}^{\pi} V_m \sin \omega t \, d(\omega t)$$

$$= \frac{V_m}{\pi} (1 + \cos\alpha)$$

so

$$V_L \text{av}(\alpha) = V_m g(\alpha)$$

and

$$I_L \text{av}(\alpha) = \frac{V_m}{R_L} g(\alpha)$$

R.M.S. values These values are found as before but with a base length π instead of 2π. Thus

$$V_L \text{rms}(\alpha) = \sqrt{\left(\frac{1}{\pi} \int_{\alpha}^{\pi} V_m^2 \sin^2 \omega t \, d(\omega t)\right)}$$

which yields

$$V_L \text{rms}(\alpha) = \frac{V_m}{\sqrt{2}} f(\alpha)$$

and

$$I_L \text{rms}(\alpha) = \frac{V_m}{\sqrt{2} R_L} f(\alpha)$$

Also

$$P_L \text{rms}(\alpha) = \frac{[V_L \text{rms}(\alpha)]^2}{R_L}$$

$$\therefore P_L \text{rms}(\alpha) = \frac{V_m^2 f^2(\alpha)}{2R_L}$$

and

$$\text{p.f.}\ (\alpha) = f(\alpha)$$

It is obvious that the same curves (*Figure 3.4*) can be used for the half-controlled bridge as for the half-wave controlled rectifier.

Worked example 3.2 The resistance of a load supplied by a half-controlled bridge increases by 20% when the power delivered is doubled. If the initial conditions are 50 V (mean d.c.) from a 240 V r.m.s. supply, calculate the firing angle for the condition at which load power is doubled.

The average value of load voltage at angle α_1 is determined by

$$V_L(\text{av})(\alpha_1) = V_m g(\alpha_1)$$

i.e.

$$50 = 240\sqrt{2}g(\alpha_1)$$

Hence

$$g(\alpha_1) = \frac{50}{240\sqrt{2}}$$
$$= \underline{0.15}$$

Using the curves of *Figure 3.4*:

$$\underline{f(\alpha_1) = 0.43}$$

Now initial load power is given by the formula

$$P_{\mathrm{L}}(\mathrm{rms})(\alpha_1) = \frac{V_{\mathrm{m}}^2 f^2(\alpha_1)}{2R_{\mathrm{L}}}$$

and, since R_{L} increases to $1.2R_{\mathrm{L}}$ at double power,

$$P_{\mathrm{L}}(\mathrm{rms})(\alpha_2) = \frac{V_{\mathrm{m}}^2 f^2(\alpha_2)}{2 \times 1.2R_{\mathrm{L}}}$$

but

$$\frac{P_{\mathrm{L}}\,\mathrm{rms}(\alpha_2)}{P_{\mathrm{L}}\,\mathrm{rms}(\alpha_1)} = 2$$

$$\therefore 2 = \frac{f^2(\alpha_2)}{1.2f^2(\alpha_1)}$$

$$\therefore f^2(\alpha_2) = 2.4f^2(\alpha_1)$$

and, since only the positive root is meaningful,

$$f(\alpha_2) = 1.55f(\alpha_1)$$
$$= 0.67$$
$$\underline{\therefore \alpha_2 \simeq 0.53\pi}$$

SINGLE-PHASE A.C. CONTROLLER

There are a number of ways of obtaining a balanced a.c. load voltage and current (i.e. with zero d.c. component). For modest load power a variety of triacs exist and since only one gate circuit is required these are generally the first choice for an a.c. controller. However, at high power a pair of thyristors appropriately connected is a more satisfactory solution.

Figure 3.6 shows the principal configurations, together with the voltage waveforms. Other configurations are possible, but these are rarely used in practice and will not be investigated here.

In circuit (a), alternate positive and negative gate pulses are supplied by a trigger module, the polarity of the gate pulse being the same as the polarity of the half-cycle in which it occurs, and at an angle α into each half-cycle. The triac conducts when triggered and the result is the load waveform shown.

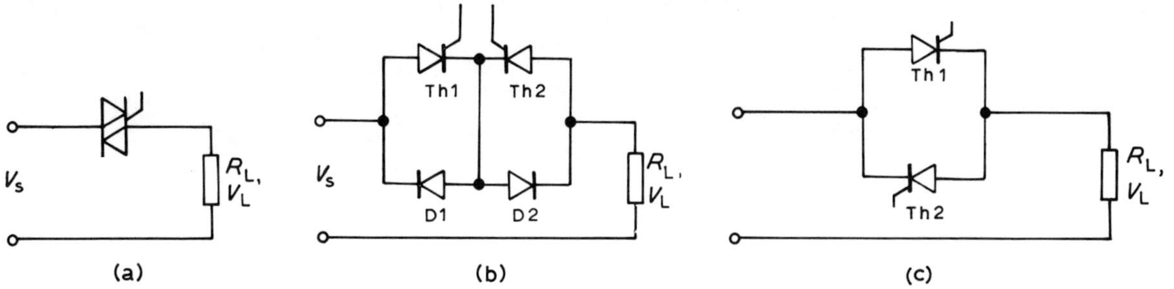

Figure 3.6 A.C. controllers: (a) triac; (b) back-to-back circuit; (c) inverse–parallel circuit; (d) load voltage waveform

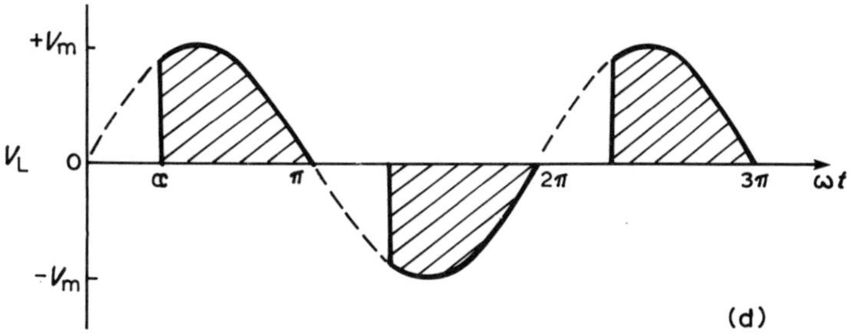

In circuit (b) when L is positive, Th1 is triggered by a positive gate pulse at angle α in the half-cycle. The conduction path which is established when Th1 turns on is from L via Th1 and D2 through the load and back to N.

In the negative half-cycle of L, N is effectively positive with respect to L and Th2 is triggered by a positive gate pulse causing current to flow from N through R_L, Th2 and D1 back to L.

The load waveform is identical with that for a triac, except for some negligible additional voltage drop across the semiconductors.

One advantage of this arrangement is that the common cathode connection for the thyristors makes the gate module design simpler.

Circuit (c) is probably the most common for high power applications. Th1 is triggered at angle α when L is in its positive half-cycle and Th2 is triggered at angle α in the other half-cycle. Because of the anode–cathode connection the gating circuits have to be electrically separate making trigger module design slightly more involved, but economy of the power components is of overriding importance at high power.

Except for the slightly different voltage drop across the thyristors, the load waveform is identical with the other two arrangements.

In *Figure 3.6* the load is shown connected to the neutral supply point. There is no electrical difference if the load is connected to the live supply point from the point of view of load waveforms.

Since the load waveform is symmetrical about zero there is no d.c. component of load voltage or current.

Formulae for the electrical conditions are developed below.

R.M.S. values Since the load waveform repeats every π radians, the r.m.s. load voltage is found from

$$V_L \text{rms}(\alpha) = \left[\frac{1}{\pi}\int_{\alpha}^{\pi} V_m^2 \sin^2 \omega t \, d(\omega t)\right]^{1/2}$$

which yields

$$V_L \text{rms}(\alpha) = \frac{V_m}{\sqrt{2}} f(\alpha)$$

and

$$I_L \text{rms}(\alpha) = \frac{V_m}{\sqrt{2} R_L} f(\alpha)$$

also

$$P_L \text{rms}(\alpha) = \frac{V_m^2 f^2(\alpha)}{2R_L}$$

further

$$\text{p.f.}\ (\alpha) = f(\alpha)$$

where $f(\alpha)$ has the same meaning as for the controlled rectifier.

Note that these expressions are identical with the r.m.s. and p.f. expressions for the half-controlled bridge circuit, and the curve of $f(\alpha)$ (*Figure 3.4*) can be used in the design of a.c. controller circuits.

Figure 3.4, then, becomes a universal design curve for single-phase thyristor circuits with resistive loads.

Hold current Any thyristor or triac when it is switched on by a gate pulse will only *latch* on if the current being conducted at the moment the gate pulse is removed is greater than a certain value called the *hold current*. In *Figure 3.7* if at angle ωt_1, the load current (and hence the thyristor current) is still below the hold value the thyristor will not continue to conduct for the rest of the half-cycle.

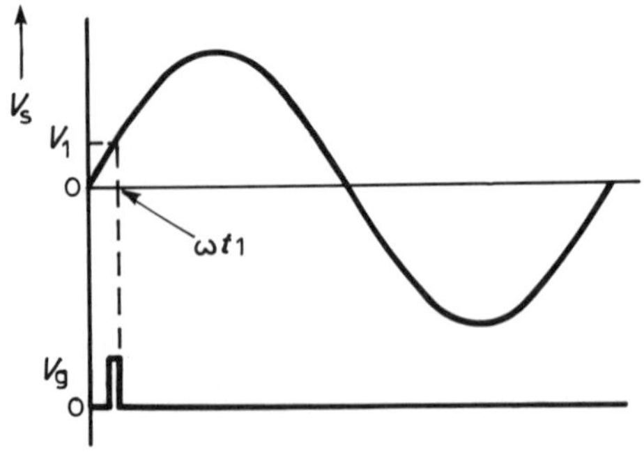

Figure 3.7 Thyristor latching

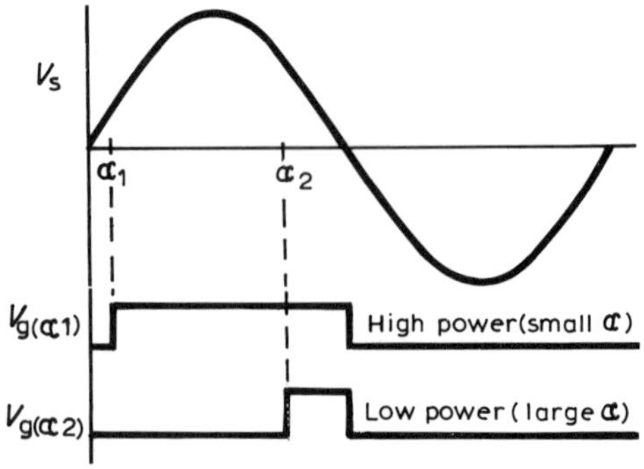

Figure 3.8 Preferred trigger waveforms

This becomes important where trigger angles are small, since it is possible for a normally working circuit at trigger angles well into the half-cycle, to *drop half cycles as α is made smaller*. In other words as α is reduced in order to increase load power, the load power is suddenly interrupted by the loss of half-cycles. In a bridge or an a.c. controller the result could be that one thyristor conducts properly and the other drops some half-cycles with very undesirable results on the load since it happens near full power.

For this reason gate pulses tend to be made rather long so that conduction is firmly established before gate drive is removed.

Some good designs maintain gate drive, once it is established, for the rest of the half-cycle, as in *Figure 3.8*.

THREE-PHASE HALF-WAVE CONTROLLED RECTIFIER

This is a circuit which is rarely used in practice because other configurations make much better use of the three phases. It is, however, a very instructive circuit. A thorough understanding of the working of this arrangement is of great help in grasping the action of the more common three-phase configurations.

Figure 3.9 shows the basic arrangement together with the voltage waveforms for low trigger angle and high trigger angle. *Note that for the circuit to operate there must be a neutral line.*

The first point to note is that if the thyristors were to be replaced by diodes conduction would pass naturally from diode to diode at points A, B, C etc., where the phases successively 'take over' the more positive potential.

These points A, B, C of *natural commutation* are the $\alpha = 0$ points for the thyristors. If $\alpha = 0$ radians the thyristors conduct as though they were diodes and the full rectified output appears across the load.

The second point to note is that there are two quite distinct ranges of conduction to consider: (a) $\pi/6 \leqslant \alpha < 5\pi/6$ (high α); and (b) $0 < \alpha < \pi/6$ (low α).

Figure 3.10 shows the two cases with just two adjacent phases shown for clarity.

Taking the high α case first, Th_R triggers when α_R is reached, and current flows through Th_R and into the load. The load voltage follows the red phase until the thyristor Th_R commutates when V_R goes negative. The next thyristor to fire is Th_Y, when α_Y is reached, and the load voltage follows the yellow phase until Th_Y commutates.

From the point of view of the phase voltages, conduction starts at angle $[(\pi/6) + \alpha]$ radians and ceases at angle π radians, and the load waveform repeats every $2\pi/3$ radians.

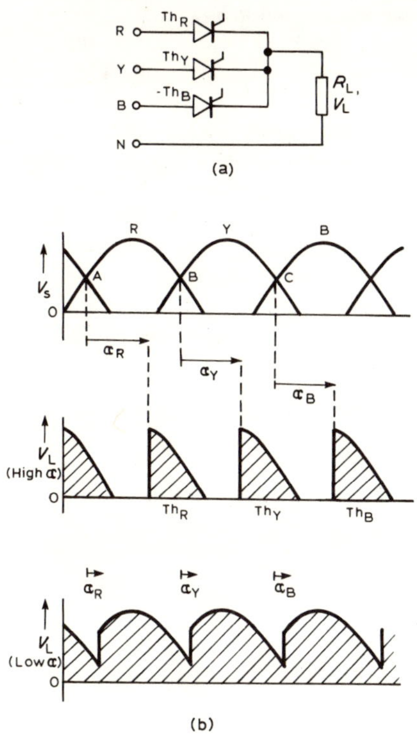

Figure 3.9 Three-phase half-wave controlled rectifier: (a) circuit diagram; (b) waveforms

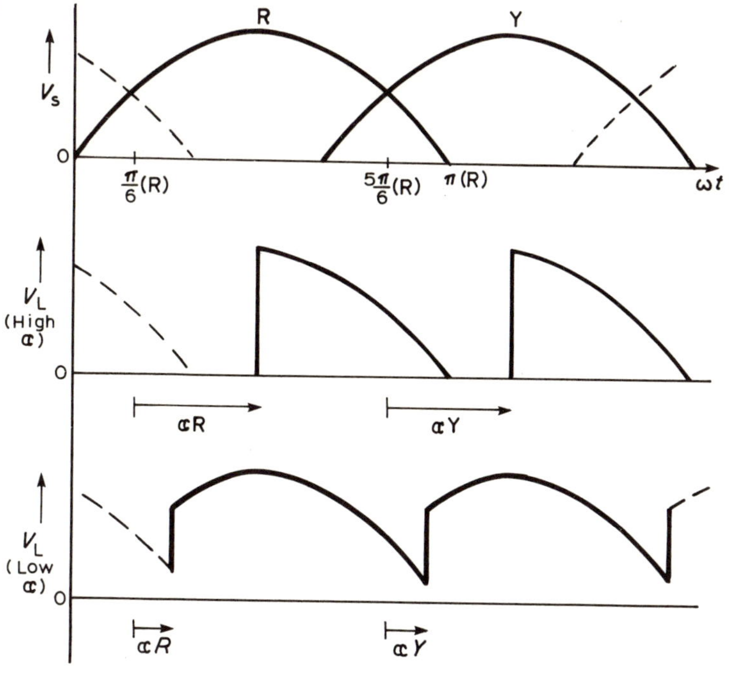

Figure 3.10 Three-phase controlled rectifier – waveform details (note: for each phase, $\alpha = 0$ coincides with $\omega t = \pi/6$ for that phase, and the maximum value of α is $\alpha = 5\pi/6$ which coincides with $\omega t = \pi$ for that phase)

This range of conduction is characterised by periods (however short) when the load voltage falls to zero. It is therefore described as the *discontinuous mode of control.*

In the case of low α (i.e. α lies between 0 and $\pi/6$ radians measured from its reference point), the red phase thyristor Th_R conducts until the next thyristor Th_Y is triggered, when this takes over from Th_R (since its anode is the more positive).

The conduction starts in any phase at angle $[(\pi/6) + \alpha]$ radians as before but terminates at angle $[(5\pi/6) + \alpha]$ when the next phase takes over via its thyristor.

At no time during this mode of conduction does the load voltage waveform fall to zero. This then is the *continuous mode of control.*

Again the load waveform repeats every $2\pi/3$ radians.

Expressions which describe the electrical condition in the load can be developed for the two cases by using the appropriate limits of integration and the base length $2\pi/3$ radians.

Average values

Discontinuous mode:

$$V_L(\text{av})(\alpha) = \frac{3}{2\pi}\int_{\pi/6+\alpha}^{\pi} V_m \sin \omega t \, d(\omega t)$$

$$= \frac{3V_m}{2\pi}\Big|-\cos \omega t\Big|_{\pi/6+\alpha}^{\pi}$$

$$= \frac{3V_m}{2\pi}\left\{1 + \cos\left(\frac{\pi}{6} + \alpha\right)\right\}$$

Thus

$$V_L \text{av}(\alpha) = 0.477 V_m\left[1 + \cos\left(\frac{\pi}{6} + \alpha\right)\right]$$

and

$$I_L \text{av}(\alpha) = \frac{0.477 V_m}{R_L}\left[1 + \cos\left(\frac{\pi}{6} + \alpha\right)\right]$$

Continuous mode:

$$V_L \text{av}(\alpha) = \frac{3}{2\pi}\int_{\pi/6+\alpha}^{5\pi/6+\alpha} V_m \sin t\omega \, d(\omega t)$$

$$= \frac{3V_m}{2\pi}\left[-\cos\left(\frac{5\pi}{6} + \alpha\right) + \cos\left(\frac{\pi}{6} + \alpha\right)\right]$$

which after expansion of the cosine terms and collecting variables together yields

$$V_L \text{av}(\alpha) = 0.827 V_m \cos\alpha$$

and

$$I_L \text{av}(\alpha) = 0.827 \frac{V_m}{R_L} \cos \alpha$$

Notice that in the discontinuous mode the general form of the expressions is similar to those for the single-phase case.

If we make

$$\delta = \left(\frac{\pi}{6} + \alpha\right)$$

then

$$V_L \,\text{av}(\alpha) = 1.5 V_m \frac{(1 + \cos\delta)}{\pi}$$

$$V_L \,\text{av}(\alpha) = 1.5 V_m g(\delta)$$

and

$$I_L \,\text{av}(\alpha) = \frac{V_m}{R_L} g(\delta)$$

Thus for this case the general curve of *Figure 3.4* can be used with $g(\delta)$ replacing $g(\alpha)$.

R.M.S. values *Discontinuous mode:*

$$V_L \,\text{rms}(\alpha) = \sqrt{\frac{3}{2\pi} \int_{\pi/6+\alpha}^{\pi} V_m^2 \sin^2 \omega t \, d(\omega t)}$$

which yields

$$V_L \,\text{rms}(\alpha) = \sqrt{\frac{3V_m^2}{4\pi} \left[\frac{5\pi}{6} - \alpha + ½ \sin 2\left(\frac{\pi}{6} + \alpha\right)\right]}$$

Again putting

$$\delta = \frac{\pi}{6} + \alpha$$

$$V_L \,\text{rms}(\delta) = \sqrt{\frac{3V_m^2}{4\pi} [\pi - \delta + ½ \sin 2\delta]}$$

The expression

$$\left(\frac{\pi - \delta + ½ \sin 2\delta}{\pi}\right)^{½}$$

is similar to $f(\alpha)$ with δ replacing α Thus for the discontinuous mode the expression reduces to

$$V_L \,\text{rms}(\delta) = 0.866 V_m f(\delta)$$

and the curve of *Figure 3.4* can be used with δ replacing α. Also

$$I_L \,\text{rms}(\alpha) = \frac{0.866 V_m f(\delta)}{R_L}$$

Continuous mode:

The limits of integration are now $[(\pi/6) + \alpha]$ and $[(5\pi/6) + \alpha]$ as before, so that

$$V_L \,\text{rms}(\alpha) = \sqrt{\frac{3}{2\pi} \int_{\pi/6+\alpha}^{5\pi/6+\alpha} V_m^2 \sin^2 \omega t \, d(\omega t)}$$

which yields

$$V_L \text{rms}(\alpha) = \frac{V_m}{\sqrt{2}} (1 + 0.413 \cos 2\alpha)^{1/2}$$

and

$$I_L \text{rms}(\alpha) = \frac{V_m}{\sqrt{2} R_L} (1 + 0.413 \cos 2\alpha)^{1/2}$$

Load power in both cases is available from

$$P_L \text{rms}(\alpha) = \frac{V_L^2 \text{rms}(\alpha)}{R_L}$$

Summarising the foregoing work, with polyphase controlled rectification there are two ranges of firing angle which yield different expressions for the load conditions. For three-phase circuits the boundary between these two ranges occurs at $\alpha = \pi/6$, or $\omega t = \pi/3$. When Figure 3.4 is used for the discontinuous case, it is important that the limit of validity is not exceeded. Figure 3.4 is valid only between $\delta = \pi/3$ and π for the case above.

Worked example 3.3 Determine, for a three-phase half-wave controlled rectifier powered by a 415 V (line) supply, the maximum average load voltage achievable in the discontinuous mode of control.

For the half-wave three-phase bridge the average load voltage in the discontinuous mode is given by

$$V_L(\text{av})(\alpha) = 1.5 V_m g(\delta)$$

where

$$\delta = \frac{\pi}{6} + \alpha$$

Here V_m is the peak phase voltage:

$$V_m = \frac{415}{\sqrt{3}} \times \sqrt{2}$$
$$= \underline{340 \text{ V}}$$

The lower limit of α for discontinuous conduction (corresponding to the maximum load voltage under this condition) is $\pi/6$, which in turn corresponds to a value of $\delta = \pi/3$.

Thus the required value is given by

$$V_L(\text{av})(\alpha) = 1.5 \times 340 \times g\left(\frac{\pi}{3}\right)$$
$$= 1.5 \times 340 \times 0.5$$
$$= \underline{255 \text{ V}}$$

THREE-PHASE HALF-CONTROLLED BRIDGE

This controlled rectifier operates without a neutral (star point return) wire and finds great application in high power control circuits. It is particularly suited to electric furnace control and the control of large industrial d.c. drives where regenerative braking is not a requirement.

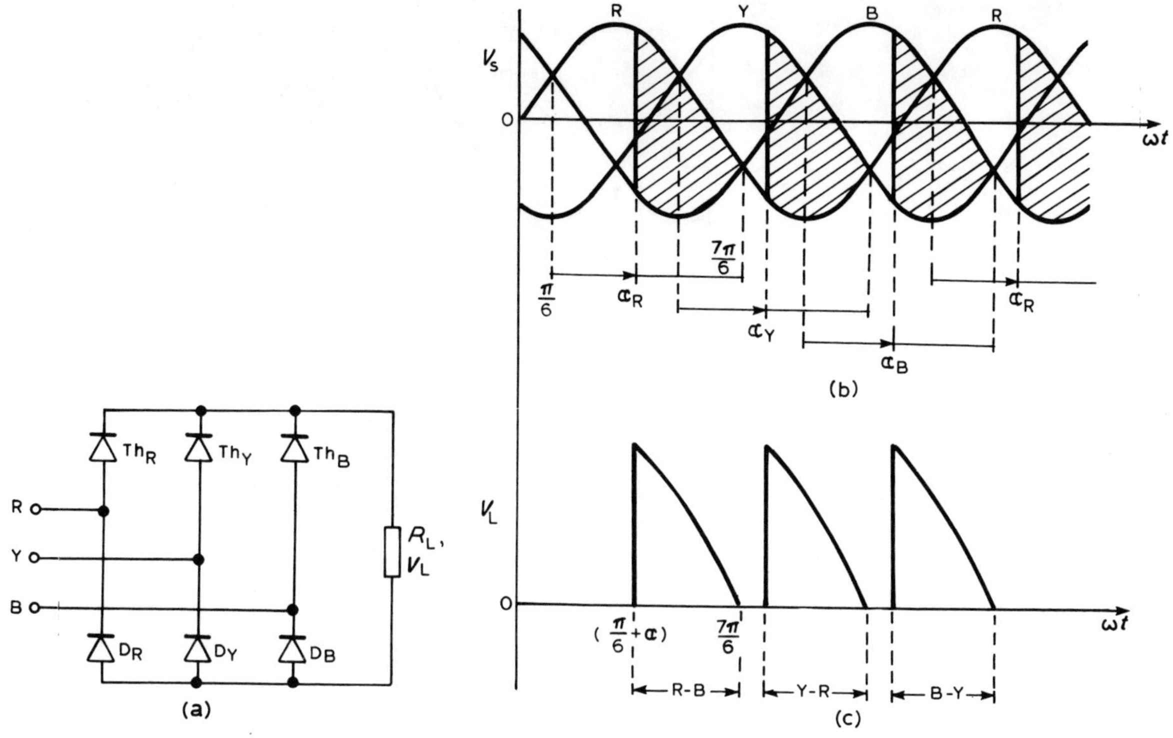

Figure 3.11 Three-phase half-controlled bridge ($\alpha \geqslant \pi/3$): (a) circuit diagram; (b) line voltage waveforms and trigger angles; (c) load voltage waveforms and trigger angles

The arrangement and voltage waveforms appear in *Figure 3.11*, for large values of trigger angle ($\alpha \geqslant \pi/3$)

To understand the operation of this bridge two points must be kept in mind. These are:

1. *When the thyristor is triggered, it will conduct through the load from its supply phase into the most negative of the other two phases.*
2. *When the supply phase of an ON thyristor becomes more negative than both the other phases the thyristor will commutate naturally.*

Referring to *Figure 3.11(a)*, the red phase thyristor Th_R is triggered at angle α_R. The most negative phase at this time is the blue phase, so current flows from the red phase, through Th_R, R_L and D_B back to the blue phase. This is indicated by the shaded region on the three-phase diagram – *Figure 3.11(b)*. From the point of view of the load, the blue phase becomes temporarily the 'reference' (i.e. the negative) potential and the difference between V_R and V_B creates the voltage V_L. The shape of this V_L is shown in *Figure 3.11(c)*.

Notice that α has a total range of 180° in this arrangement. Th_R commutates when the red phase becomes the most negative, and the next thyristor to be triggered is Th_Y.

Figure 3.11(c) also indicates the phases carrying current for various parts of the load waveform, and inspection of *Figure 3.11(b)* reveals that, from the point of view of the more positive conducting phase, conduction commences at angle $[(\pi/6) + \alpha]$ radians and ceases at $7\pi/6$ radians.

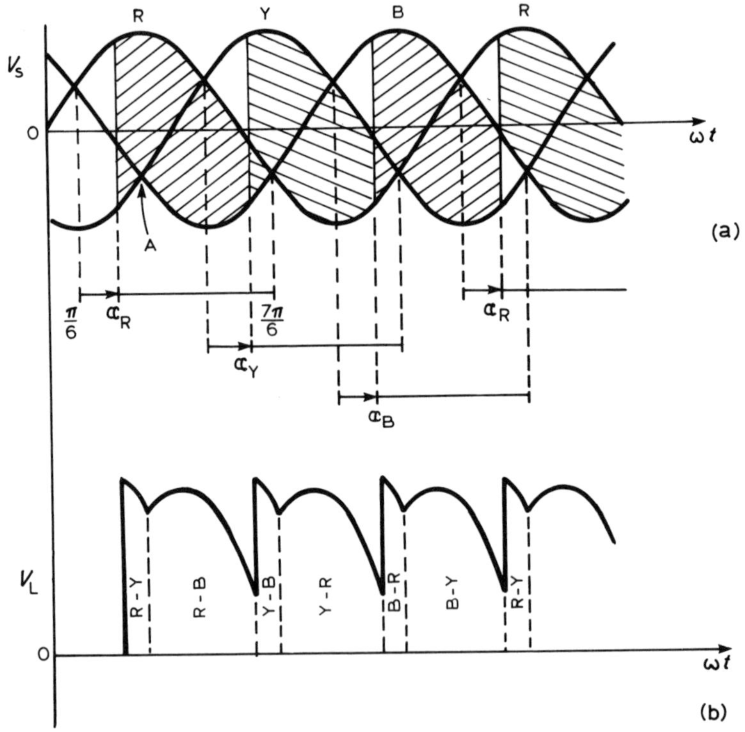

Figure 3.12 Three-phase half-controlled bridge ($\alpha < \pi/3$): (a) line voltages and trigger angles; (b) load voltage

For the case where $\alpha < \pi/3$ two phases become, in turn, the most negative phase and return current flows first into one and then into the other. *Figure 3.12* illustrates this effect.

Starting with Th_R being triggered at angle α_R, current flows from the red phase via the load and back into the yellow phase. At point A, phase B becomes more negative than phase Y so the blue phase takes over the role of the return phase.

Inspecting *Figure 3.12(a)* it is evident that the yellow phase thyristor Th_Y is triggered before the load voltage has fallen to zero. When Th_Y is triggered Th_R turns off. This is because when Th_Y turns on, the yellow phase is more positive than the red phase so the cathode of Th_R is carried more positive than its anode.

Figure 3.12(b) shows the load voltage waveform and it is seen that conduction is continuous.

The average load voltage for both continuous and discontinuous cases is

$$V_L(\text{av})(\alpha) = \frac{3\sqrt{3}}{2} V_m g(\alpha)$$

It should be noted that V_m is the peak phase voltage.

THREE-PHASE FULLY CONTROLLED BRIDGE

One of the problems resulting from using a half-controlled three-phase bridge is that the ripple frequency is six times the supply frequency at low values of α (see *Figure 3.12*) and three times the supply frequency at high values of α (see *Figure 3.11*). The fully controlled bridge supplies a d.c. output with a ripple frequency of six times the supply frequency

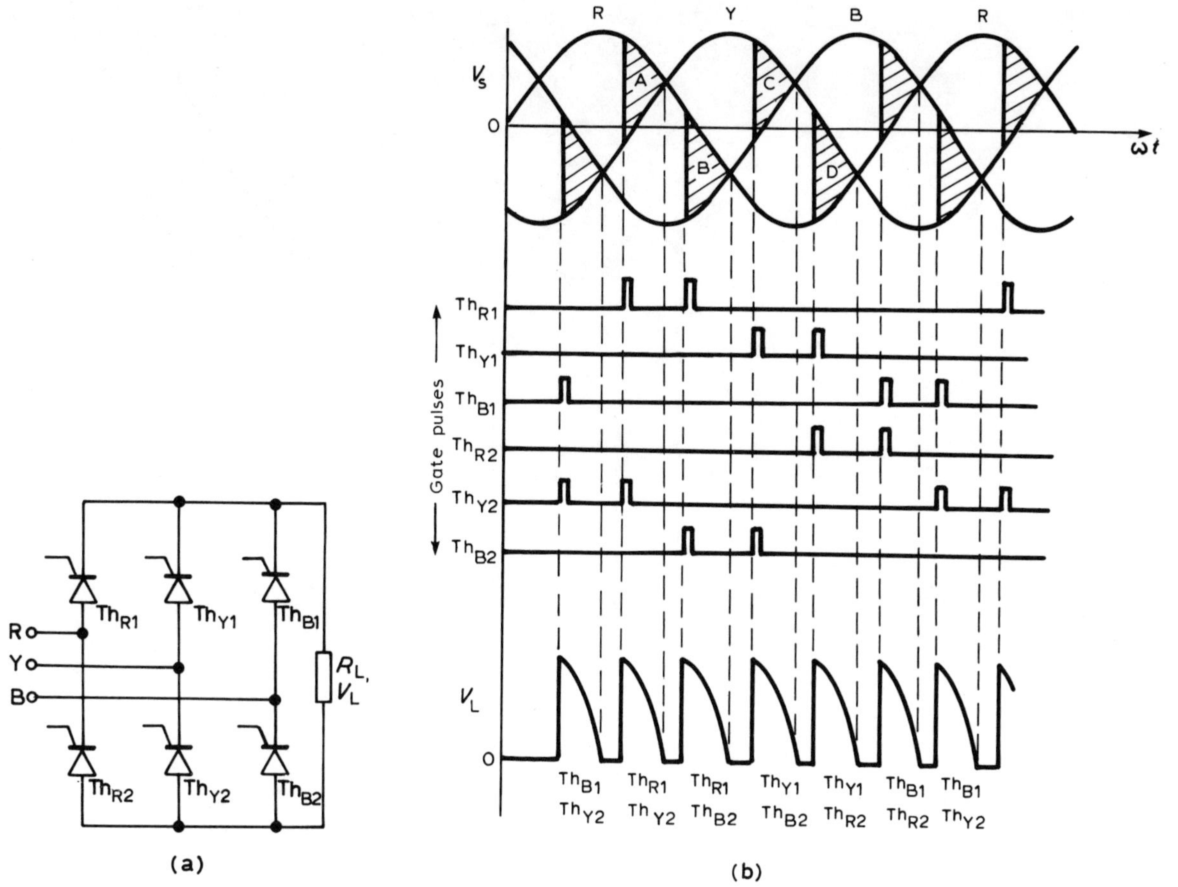

Figure 3.13 Three-phase fully controlled bridge: (a) circuit diagram; (b) waveforms (large α)

at all values of α. This makes the design of smoothing circuitry simpler. In addition a fully controlled bridge provides the means for *regeneration* (see Chapter 4). *Figure 3.13(a)* shows the circuit diagram for a fully controlled bridge with a resistive load, and *Figure 3.13(b)* shows the waveforms obtained for a high value of α.

It is necessary to arrange the triggering so that each of the thyristors Th_{R1}, Th_{Y1}, Th_{B1} conducts in turn into the other two phases. Each thyristor receives a train of double gate pulses per cycle, spaced $\pi/3$ apart. For example Th_{R1} has two opportunities to conduct, first into the yellow phase via Th_{Y2} (shaded area A in *Figure 3.13(b)*), then into the blue phase via Th_{B2} (shaded area B). Thus Th_{R1} and Th_{Y2} are triggered together, then Th_{R1} and Th_{B2}. The next pair is Th_{Y1} with Th_{B2} followed by Th_Y with Th_{R2} (shaded areas C and D).

By inspecting the waveforms it becomes clear that the trigger sequence is

R_1	Y_2
R_1	B_2
Y_1	B_2
Y_1	R_2
B_1	R_2
B_1	Y_2

If the whole of the trigger pulse waveforms are shifted to the left, the firing of thyristor pairs occurs earlier in the cycle. *Figure 3.14* shows the result for a small value of α, and also shows the thyristor pair in conduction at any time.

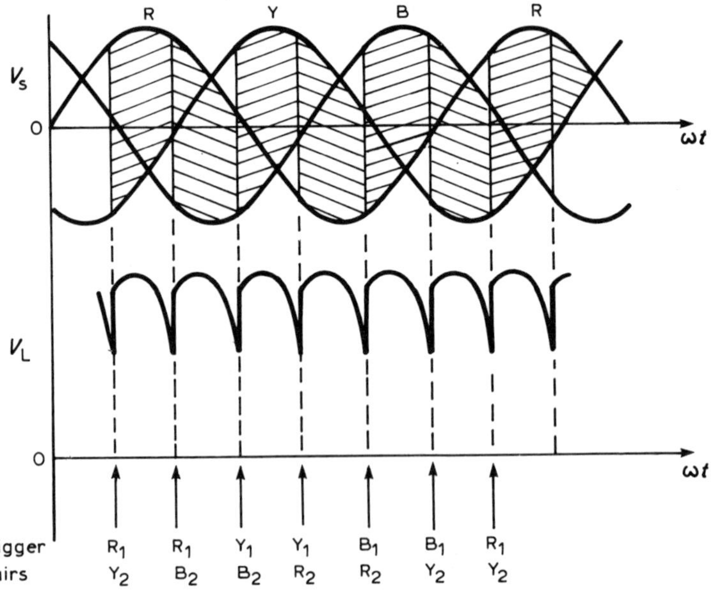

Figure 3.14 Three-phase fully controlled bridge waveforms (small α)

PROBLEMS FOR CHAPTER 3

(1) A single-phase a.c. controller is to deliver 2 kW into a 5 Ω load from a 240 V_{rms} supply. Determine the trigger angle and the power factor presented to the supply.

Answer: $\alpha = 0.68\pi$; p.f. = 0.416

(2) A half-controlled bridge is designed to provide 100 V average across a resistive load of 20 Ω, from a 240 V main supply. Under a fault condition in which gate pulses are not available to one thyristor, calculate the load power.

Answer: 640 W

(3) An electric kiln is controlled by a triac from a 240 V supply. At full power (represented by trigger angle $\alpha = \pi/5$) the load power is 2 kW. Assuming the resistance of the element varies with load power according to the equation

$$R_L = R_M[1 - \gamma(\Delta P_L)^{1/2}]$$

where R_L = element resistance
R_M = resistance at full power
γ = 0.3

$$\Delta P_L = \frac{\text{Original power} - \text{Final power}}{\text{Original power}}$$

determine the r.m.s. load voltage and current at half-power.

Answer: $V_L(\text{rms})(\alpha) = 144$ V; $I_L(\text{rms})(\alpha) = 6.95$ A

(4) A half-wave controlled rectifier consisting of a single thyristor is designed to deliver 100 W into a 10 Ω load when run from a 50 V r.m.s. supply. The thyristor can pass a maximum mean current of 6 A. Determine:

(a) The trigger angle for the design load power.
(b) The largest value of r.m.s. supply voltage which will not result in excessive mean thyristor current at this trigger angle.

Answer: (a) 0.33π; (b) 178 V(r.m.s.)

(5) Sketch the voltage waveform across a triac which is working at a trigger angle of 0.5π.

(6) The back-to-back a.c. controller in *Figure 3.6* is operating at a trigger angle of 0.5π. Sketch the load voltage waveform under the fault condition that diode D_2 is a permanent short-circuit. What effect will this have on the r.m.s. and d.c. conditions of the load?

Determine by integration the mean d.c. load voltage under these conditions, for a 240 V mains supply.

Answer: −54 V

(7) A three-phase half-controlled bridge supplying a resistive load is triggered at angle $\alpha = \pi/2$ radians. Determine the conduction angle per thyristor. If α is reduced to $\pi/9$ radians determine the new conduction angle per thyristor. Sketch the load waveform in each case.

Answer: $\alpha = \pi/3, 2\pi/3$

(8) Explain the process by which load current is transferred from diode to diode in a half-controlled bridge operating under continuous conduction conditions.

(9) A half-controlled three-phase bridge is supplied from a 415 $V_{r.m.s.}$ (line) source. Determine the required trigger angle for an average load voltage of 200 V.

Answer: $\alpha = 0.5\pi$

4 Power control with a.c. supplies—II

Many thyristor applications involve the control of power delivered to an inductive–resistive or a motor load. In this chapter the basic concepts of this type of control are examined, mainly descriptively.

SINGLE-PHASE CONTROLLED RECTIFIER WITH COMPLEX LOAD

Figure 4.1(a) shows the basic circuit for single-phase half-wave control. The load is shown connected in the cathode lead of the thyristor, but it makes no electrical difference if it is connected in the anode lead. The firing angle is assumed to be α, and the waveforms obtained are shown in *Figure 4.1(b)*. If the diode D, called a *flywheel diode*, is connected as indicated, it has a profound effect on the waveforms obtained for both load voltage and load current. This is also shown in *Figure 4.1(b)*.

Consider first the case where the flywheel diode D is not connected. This is fairly rare in controlled rectifier circuits, but it is very instructive to examine the waveforms because it helps to fix ideas about the operation of a.c.–a.c. converters with inductive loads, for example with a triac.

In *Figure 4.1(b)* the thyristor fires at angle α. The load voltage V_L immediately switches to the supply voltage (minus the small drop across the thyristor), and then V_L follows V_s. The load current I_L begins to rise as soon as the thyristor switches on, but because of the time constant

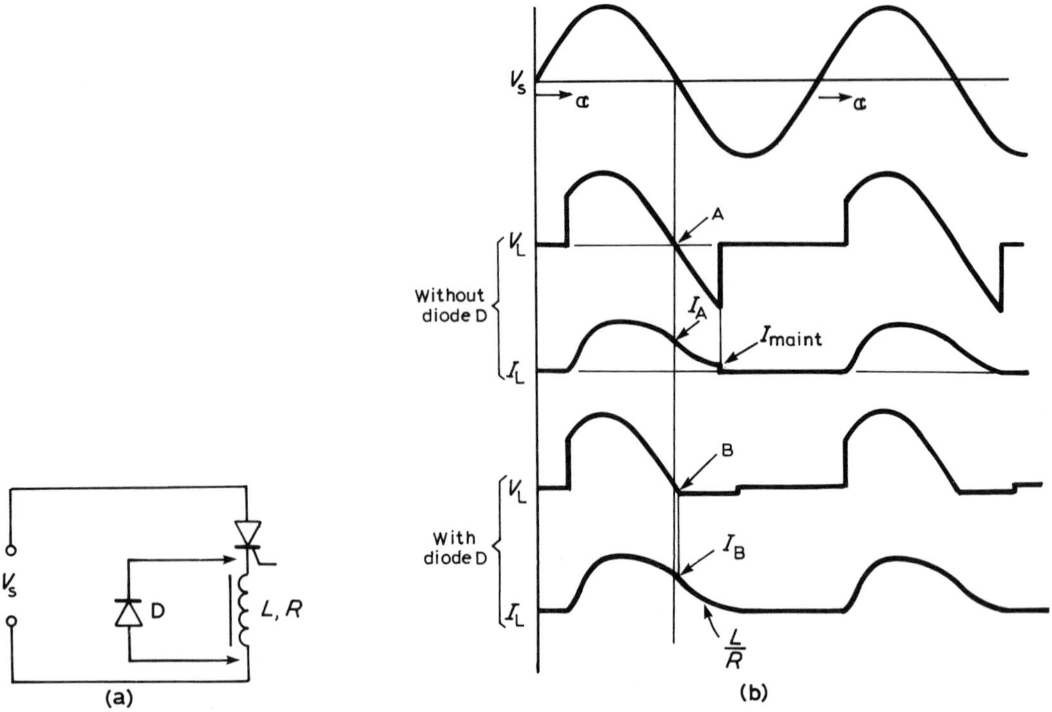

Figure 4.1 Thyristor with complex load: (a) circuit diagram; (b) waveforms

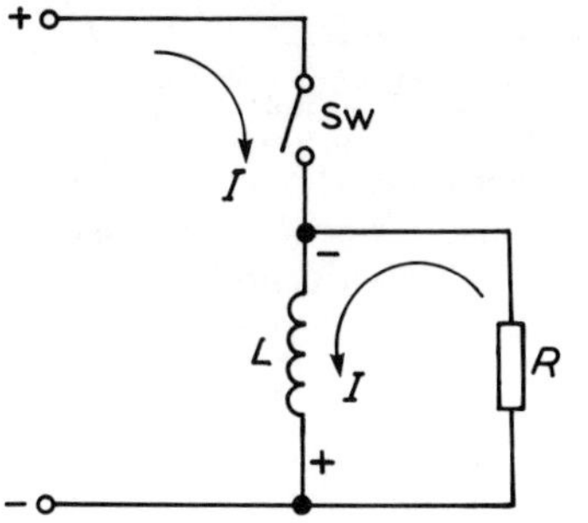

Figure 4.2 Generation of internal e.m.f.

L/R of the load it rises according to a combination of an exponential and a sine wave. The result is a current waveform having the well-known lopsided appearance shown. At point A on the V_L waveform the supply voltage V_s goes negative and if the load were purely resistive the thyristor would turn off. However, at this time the load (and thyristor) current is at a fairly high value, shown as I_A on the waveform. The inductance of the load will attempt to keep current flowing in the same direction and to do this it generates an internal e.m.f. which keeps the thyristor cathode more negative than its anode. This point is best understood by reference to *Figure 4.2* which shows an inductor, a switch and a conducting path provided by R placed across the coil.

When Sw is opened after being closed long enough to establish a current I in the inductor, the coil generates an internal e.m.f. with the polarity shown in order to maintain I in the same direction through the coil. Notice that the end of the coil near the switch is the more negative.

Referring back to *Figure 4.1(b)*, I_L continues to flow as V_s and V_L go negative, until the point is reached when I_L is equal to the *maintaining current* I_{maint} for the thyristor. At this point the thyristor switches off and V_L and I_L become zero. The process repeats in the next positive half-cycle of V_s.

When the flywheel diode D is connected the waveforms change considerably. At first, when the conduction starts, the same results are obtained. However, when V_s goes negative the inductor generates its internal e.m.f. as before, but this has the effect of forward biasing the diode. At point B on the V_L waveform the thyristor cathode has gone negative by about 1 V (the turn-on potential of the diode) when D conducts. The load current at this point is I_B, and this is switched into the diode. Since the thyristor anode continues to move negatively with V_s, and since its cathode is held at −1 V by D, the thyristor turns off. The current in the load then decays to zero from a value I_B on a time constant L/R. During the decay the thyristor cathode stays at −1 V, and when I_L becomes zero V_L returns to zero. Strictly, the diode forward resistance contributes to the time constant of the decay but usually this can be ignored.

Diode D is called a *flywheel diode* because it takes over I_L and allows a slow run-down of current rather like the slow run-down of angular velocity of a flywheel when the drive is removed.

The flywheel diode connection is very common in thyristor rectifier circuits, for example where the thyristor is controlling the field current of a shunt d.c. motor.

Practical circuits – effect of time constant

Generally speaking the time constant τ of an inductive–resistive load will be of the same order of magnitude as, or larger than, the period of the supply waveform. For example, a shunt field of a d.c. motor may have a time constant of 10–50 ms. This means that I_L may not decay completely to zero through the flywheel diode D in *Figure 4.1(a)*, before the thyristor fires again. There will be a period of time lasting for several cycles of supply when the mean load current is increasing, and this is shown in *Figure 4.3*.

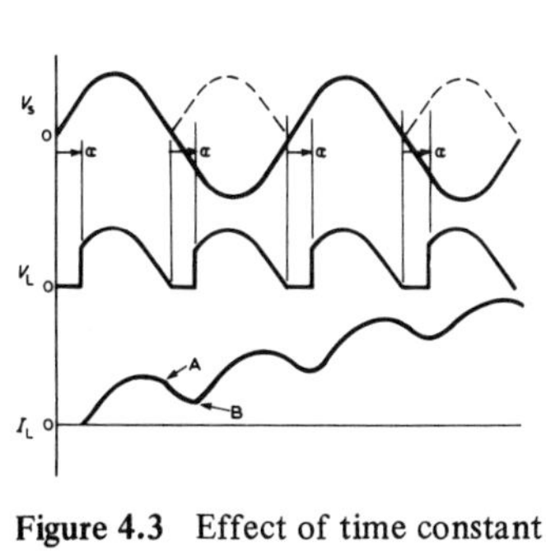

Figure 4.3 Effect of time constant

In this case a half-controlled bridge (see Chapter 3) is assumed to be controlling the voltage and current in the load. In the first half-cycle I_L flows according to the normal lop-sided waveform. At point A in *Figure 4.3* the flywheel diode conducts and I_L starts to decay. If the time constant is long, a substantial current is still flowing in the load at point B, when the bridge again supplies current. The load current rises again in the second half-cycle, and in this way the average value of I_L increases. Eventually a steady condition is reached when the energy supplied by the bridge in any half-cycle is equal to the energy lost during the decay period. Under steady conditions, provided the time constant is as described above, a convenient approximation which is often made is to assume that the increase in I_L during a half-cycle is negligible, and that the decay is also negligible. *Figure 4.4(b)* shows the approximate

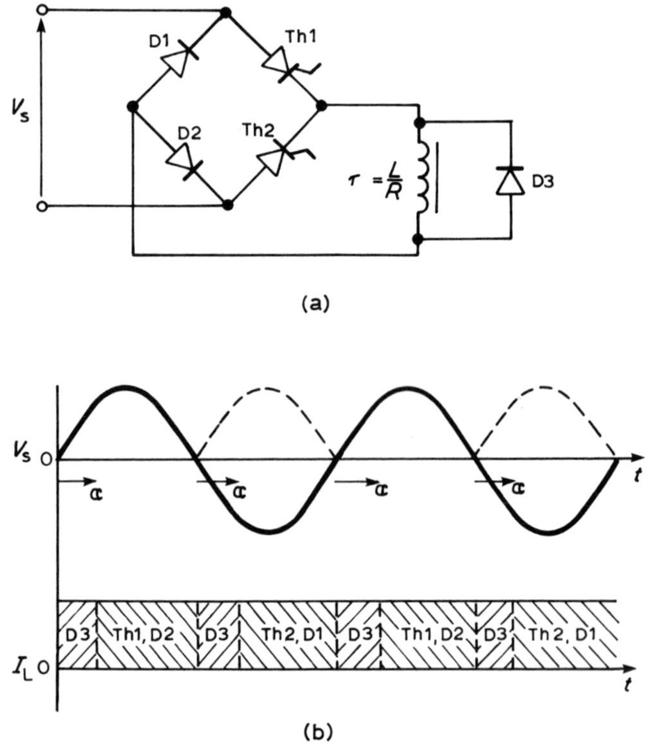

Figure 4.4 Effect of long time constant: (a) half-controlled bridge; (b) current waveform (large τ)

result for the bridge shown in *Figure 4.4(a)*. In *Figure 4.4(b)* the conducting elements are also labelled under the appropriate parts of the waveform.

Worked example 4.1 A half-controlled bridge supplies field current to a d.c. motor field coil of inductance 0.1 H and resistance 10 Ω. Determine the firing angle α required for a mean current of 5 A from a mains supply of 240 V r.m.s.

For the field coil:

$$\tau = \frac{L}{R} = \frac{0.1}{10} = \underline{0.01\ \text{s}}$$

This is of the same order of magnitude as the half-cycle period of the 50 Hz supply and therefore the constant current approximation can be made.

Since the current is steady, the mean d.c. voltage across the coil is

$$\begin{aligned} V_L &= I_L R \\ &= 5 \times 10 \\ &= 50 \text{ V} \end{aligned}$$

But this must equate to $V_{L(av)(\alpha)}$, the mean bridge output voltage (see Chapter 3). Thus, from Chapter 3,

$$\begin{aligned} V_{L(av)(\alpha)} &= \frac{V_m}{\pi}(1 + \cos\alpha) \\ &= V_m\, g(\alpha) \end{aligned}$$

where $g(\alpha)$ is defined in *Figure 3.4.* Thus

$$\begin{aligned} g(\alpha) &= \frac{50}{240\sqrt{2}} \\ &= 0.147 \end{aligned}$$

and using *Figure 3.4*

$$\underline{\alpha = 0.675\pi \text{ rad}}$$

It is worth reflecting for a moment on what constitutes a 'long' time constant. The important feature is the amount of time during which the flywheel diode conducts and I_L decays. This will depend on whether the circuit involves a full-wave (bridge) or a half-wave controlled rectifier.

For the full-wave case the decay period is less than a half-cycle, whereas for the half-wave case the decay period is greater than a half-cycle. Since decay to zero would occupy about 5 time constants, τ can be considered 'long' if it is equal to or greater than a half-cycle of supply, i.e.

$$\tau \geqslant \frac{\pi}{\omega}$$

where ω = frequency of supply (rad/s).

Triggering with inductive loads

If the trigger circuit is one which provides short duration pulses of, say, 100 μs such as are suitable for resistive loads, it is quite likely that the thyristors will remain in the off condition when the load is inductive. This is because of the much slower rate of rise of load current, so that the latching current for the thyristors will not be attained during the short gate pulse time. To overcome this problem trigger circuits are designed to provide long gate pulses, lasting typically up to the natural commutation point in the supply waveform. In some cases it may be necessary to fire the thyristors earlier in the half-cycle and adjust the firing angle as the load current builds (usually automatically, using a current sensing element in the trigger module).

SINGLE-PHASE A.C.–A.C. CONVERTER WITH COMPLEX LOAD

All of the circuits shown earlier in *Figure 3.6* may be used with inductive–resistive loads to obtain an a.c. output.

The load current will consist of positive and negative waveforms having the same lop-sided shape as for the rectifier above, *provided the trigger* *angle α is greater than the load phase angle ϕ.* The waveform is *discontinuous* under these conditions (i.e. it has periods of zero value). *If the trigger angle α is made equal to the load phase angle ϕ, the load current becomes sinusoidal* – in fact the circuit behaves as though the thyristors were not there at all. The circuit operation is not quite as simple as that, however. *Figure 4.5* shows the waveforms and circuit diagrams for a back-to-back thyristor pair.

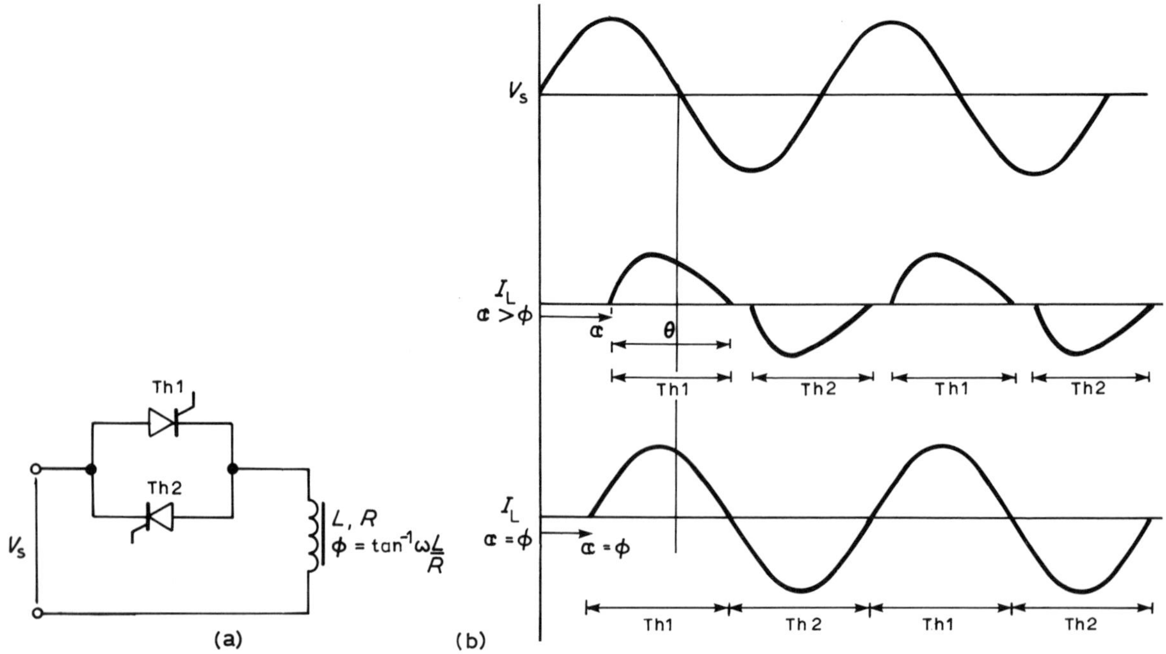

Figure 4.5 A.C. converter with complex load; (a) back-to-back thyristor pair with complex load; (b) waveforms

Referring to *Figure 4.5*, with $\alpha > \phi$, each thyristor conducts in turn well into its respective negative half-cycle of supply voltage, but its conduction angle θ is less than 180°. What happens is that when the supply voltage passes through zero there is still a substantial load current. The inductor generates an internal e.m.f. which keeps the conducting thyristor forward biased until the current falls below the maintaining value for the thyristor.

As α is made smaller the conduction angle approaches 180° and the current waveform becomes progressively more sinusoidal and at $\alpha = \phi$ the waveform is a sinusoid.

Figure 4.5(b) indicates which thyristors are conducting for the various parts of the waveform.

It is important to note what happens if α is made less than ϕ.

If the trigger is a short duration pulse Th1 will conduct (say) and continue to conduct for more than 180°. In the meantime the short trigger pulse has come and gone unnoticed by Th2 (since Th2 is reverse biased whilst Th1 conducts). Th2 therefore does not trigger and the net result is a rectified output, the polarity of which depends on which thyristor is triggered first.

This situation can be resolved by using long duration trigger pulses so that the thyristors can start conducting at $\alpha = \phi$ even though the pulse is applied earlier.

There is not space in this book to investigate the design procedures for a.c.–a.c. converters with complex loads, and the reader is referred to *Power Engineering Using Thyristors*, vol. I, published by Mullard Technical Publications, for a full treatment of the subject. *Table 4.1* summarises the main electrical conditions for inverse-parallel and back-to-back circuit configurations with an inductive/resistive load, for trigger angle α in the range $\phi \leqslant \alpha \leqslant \pi$. In the table, V_s is the r.m.s. supply voltage and the other parameters have the significance indicated in *Figure 4.5(b).*

Table 4.1 *Main formulae for a.c. converter with inductive–resistive loads*

Quantity	*Expression*
R.M.S. load voltage	$V_L(\text{rms})(\alpha) = V_S\left[\frac{\theta - \sin\theta\cos(2\alpha+\theta)}{\pi}\right]$
R.M.S. load current	$I_L(\text{rms})(\alpha) = \frac{V_S}{Z_L}\left[\frac{\theta}{\pi} - \frac{\sin\theta}{\pi\cos\phi}\cos(2\alpha+\theta+\phi)\right]^{1/2}$
R.M.S. load power	$P_L(\text{rms})(\alpha) = \frac{V_S^2\cos^2\phi}{R_L}\left[\frac{\theta}{\pi} - \frac{\sin\theta}{\pi\cos\phi}\cos(2\alpha+\theta+\phi)\right]^{1/2}$
Power factor presented to supply	$\text{p.f.}(\alpha) = \cos\phi\left[\frac{\theta}{\pi} - \frac{\sin\theta}{\pi\cos\phi}\cos(2\alpha+\theta+\phi)\right]^{1/2}$

Worked example 4.2 A back-to-back thyristor pair is used to control the current in an inductive–resistive load having a time constant of 10 ms. The firing angle α is $\pi/6$ radians and the supply frequency is 50 Hz. Determine the minimum length of trigger pulse to ensure a symmetrical load current waveform.

To guard against rectification the trigger pulse must endure at least until a point corresponding to the load phase angle ϕ.

In this case $\phi = \tan^{-1} 2\pi f\tau$, where τ is the time constant:

$$\begin{aligned}\phi &= \tan^{-1} 314.2 \cdot 10 \times 10^{-3}\\ &= \tan^{-1} 3.142\\ &= 0.4\pi\end{aligned}$$

$$\begin{aligned}\text{Minimum pulse length} &= \phi - \alpha:\\ &= 0.4\pi - 0.167\pi\\ &= 0.233\pi\\ &\doteqdot \underline{2.33\text{ ms}}\end{aligned}$$

D.C. MOTOR CONTROL USING A.C. SUPPLIES

A very wide range of engineering systems require a rotating shaft which can deliver a controllable speed, torque and power to some part of the system. There are many ways of achieving this, among them controllable a.c. commutator or slipring motors, magnetic, hydrauilic or mechanical coupling systems, and d.c. motors with controllable supplies.

For many applications the d.c. motor with controllable armature and field supplies provides the most satisfactory solution and in this section the mechanism of thyristor control of shunt d.c. motors will be investigated.

There are basically three methods of controlling d.c. shunt motors, each with its own output characteristics.

1. *Constant field with controlled armature voltage.* This results in a constant output torque over the entire range of speed, and a shaft power which increases linearly with speed. *Figure 4.6(a)* illustrates this condition.
2. *Constant armature supply with controlled field current.* This results in a practically constant shaft power with speed, and a reducing torque–speed characteristic. This is illustrated in *Figure 4.6(b)*.
3. *Controlled field current with controlled armature voltage*. This results in a composite characteristic derived from the above two cases.

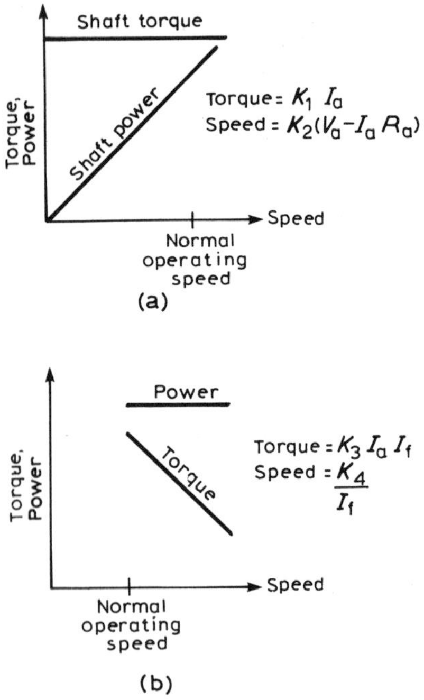

Figure 4.6 D.C. shunt motor characteristics: (a) constant I_f, controlled V_a; (b) constant V_a, controlled I_f

Referring to *Figure 4.6(a)*, the shaft torque and power are shown somewhat idealised, as functions of shaft speed. The normal operating speed represents the condition for the normal rated armature and field supplies. The speed is controllable from zero up to the normal rated conditions.

In *Figure 4.6(b)* the normal operating speed is shown as the starting point since normally this represents the condition for maximum field strength. I_f is reduced from this value to secure control, resulting in speeds above the normal rated value.

The area of principal interest in this book is concerned with fixed field, controlled armature conditions, and this is now examined. The circuit diagram for the basic system is shown in *Figure 4.7*, which shows a single-phase half-controlled bridge supplying the armature of a d.c. motor with a separately excited field.

The motor armature, while it is rotating, generates a steady back e.m.f., shown as E_b in *Figure 4.7*. The level of the back e.m.f. depends on the

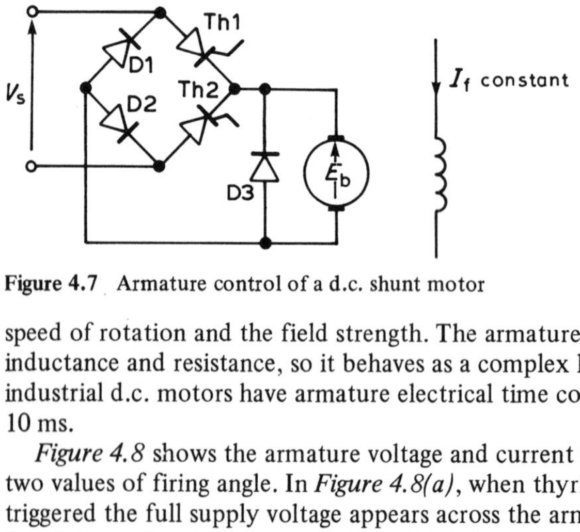

Figure 4.7 Armature control of a d.c. shunt motor

speed of rotation and the field strength. The armature also possesses inductance and resistance, so it behaves as a complex load. Indeed most industrial d.c. motors have armature electrical time constants of at least 10 ms.

Figure 4.8 shows the armature voltage and current waveforms for two values of firing angle. In *Figure 4.8(a)*, when thyristor Th1 is triggered the full supply voltage appears across the armature. V_a therefore follows the supply voltage. At point A the supply goes below the back e.m.f. level E_b generated by the motor. If there were no armature inductance Th1 would switch off at this point because its anode would become more negative than its cathode. However, a fair amount of armature inductance does exist, so Th1 is kept conducting in the same manner as for a static inductive load. V_a therefore is forced to follow the supply voltage until this goes negative at point B. Here the inductive current is still flowing, but D3 takes this over allowing Th1 to switch off. When the inductive current, now circulating in the armature and D3, has at last decayed to zero (point C in *Figure 4.8(a)*), the diode D3 switches off and the armature voltage returns to its normal back e.m.f. value of E_b.

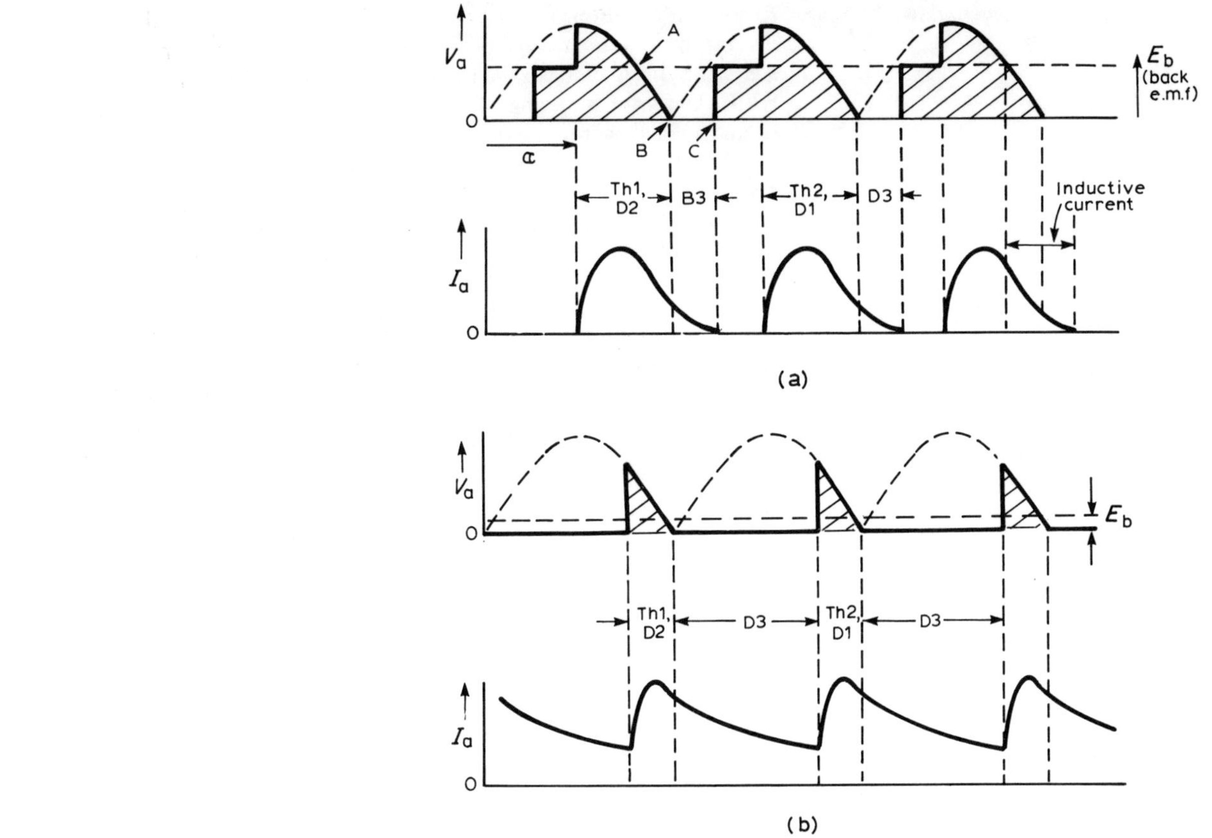

Figure 4.8 Armature voltage and current waveforms for armature control: (a) discontinuous armature current; (b) continuous armature current

The armature current consists of a discontinuous series of pulses of the form shown in *Figure 4.8(a)*. During the periods of conduction of the thyristors the current is determined by the supply waveform, and during the period of flywheel diode D3 conduction the current decays exponentially to zero at a rate which depends on the time constant of the armature and flywheel diode combination.

Figure 4.8(b) shows the effect of a low shaft speed for a motor with a fairly long armature time constant (for example under heavy load). During the period of flywheel diode conduction the armature current decays as before, but because of the low value of E_b the starting point for this decaying current is higher than before.

Under these circumstances, if the armature inductance is high enough, the current in D3 has not decayed to zero by the time the next thyristor conduction period starts. What now happens is that the armature current rises from a non-zero value. After a few cycles of steady operation the system settles and the waveforms in *Figure 4.8(b)* are obtained. The armature current is now continuous. Speed control of the motor is obtained by controlling the trigger angle α. Referring to *Figure 4.8*, if $\alpha = 180°$, the thyristors do not conduct and the motor shaft is stationary. As α is progressively reduced armature current flows and the shaft starts to rotate, causing the back e.m.f. E_b to rise. If the motor is then loaded the shaft speed and back e.m.f. fall resulting in greater armature current flowing.

Regenerative braking

In some applications it is necessary to stop, and sometimes reverse, a d.c. motor in a short space of time. One of the most effective ways of achieving this is to make the motor deliver power back into the supply. While the motor is driving its load in the forward direction, a good deal of mechanical energy is stored in the inertia of the load. It is this energy which is *re-converted into electrical energy* and forced back into the main supply during regenerative braking. There needs to be some differences in circuitry for regeneration to take place. In the case of the half-controlled bridge, as in *Figure 4.8*, when the supply goes negative the conducting thyristor and diode turn off and the flywheel diode takes over the armature current. If the flywheel diode were to be removed the conducting thyristor would continue to conduct, but its partner diode would turn off and armature current would switch to the other diode thus, for example if initially Th1 and D2 are conducting, when V_a goes negative Th1 and D1 carry the decaying armature current. In either case the motor is automatically isolated from the supply at the zero point. For regeneration this must be prevented, and a fully controlled bridge is used.

A simple system for regenerative braking is shown in *Figure 4.9*.

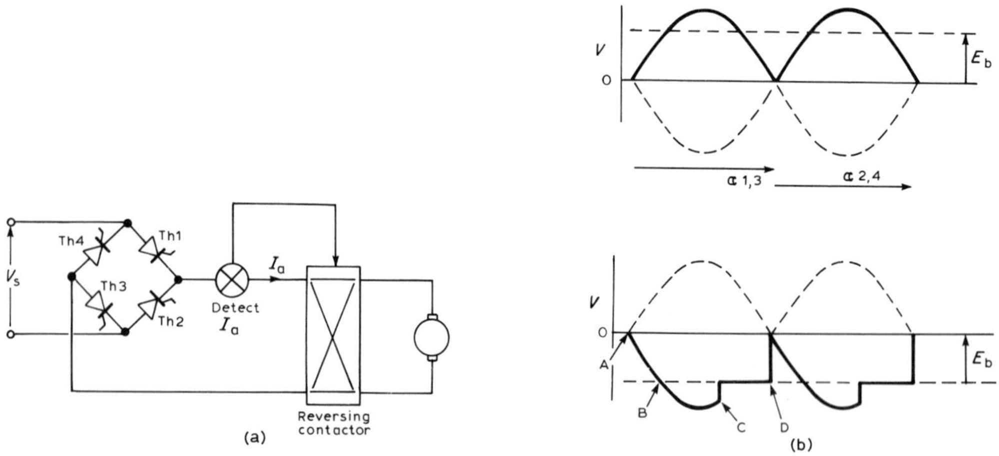

Figure 4.9 Principle of regenerative braking: (a) circuit diagram; (b) waveforms

Assume that the motor is running and generating a back e.m.f. E_b. The firing angles are now retarded to 180° so that no power is supplied and the motor is free running, still generating E_b. As soon as the thyristors cease conduction the current from the supply, I_a, becomes zero, and this condition is detected and made to operate a reversing contactor. The result of this is to connect E_b the reverse way round to the bridge terminals. *In effect, the thyristors, still with firing angles of 180°, can now conduct.* For example at point A in *Figure 4.9(b)*, thyristors Th1 and Th3 are fired. Because V_s at this point is more positive than E_b, conduction occurs with V_s going negative. At point B, V_s reaches the value of E_b, and Th1 and Th3 continue to conduct due to the inductance of the armature. At point C the inductive current has decayed to zero, Th1 and Th3 turn off and the voltage level returns to the E_b value. At point D, Th2 and Th4 are fired. Under these conditions the motor is generating into the supply and its speed (and back e.m.f.) rapidly falls to zero. If α is now advanced from 180°, the motor rotates in the opposite direction.

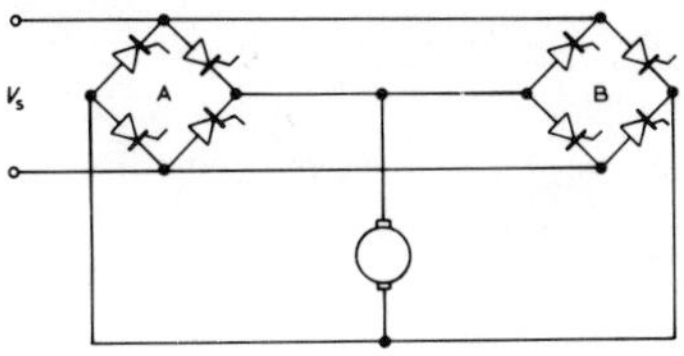

Figure 4.10 Anti-parallel bridge for regenerative braking and reversal

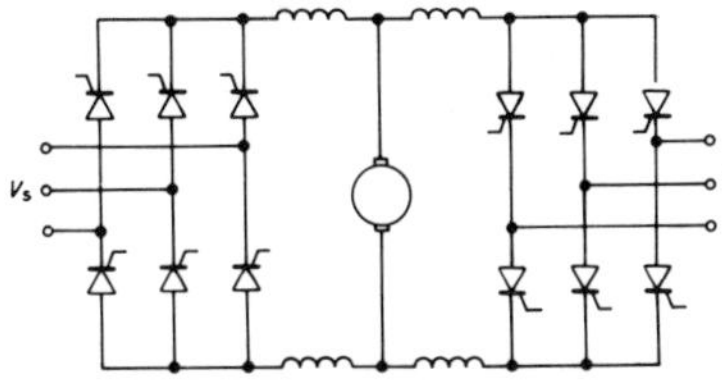

Figure 4.11 Three-phase anti-parallel bridge

A slightly more sophisticated arrangement uses the so-called anti-parallel bridge connection as in *Figure 4.10.*

For forward running, bridge A is fired and bridge B receives no gate signals at all. For reverse running bridge B is fired and bridge A receives no gate signals.

If bridge A is controlling the motor and regenerative braking is required, the gate signals are removed from bridge A and applied on bridge B at 180° firing angle. This has the same effect as the reversing contactor in *Figure 4.9*, and the motor regenerates into the main supply via bridge B. Some precautions have to be taken with this kind of arrangement, the main one being to guard against accidental firing of both bridges together. This places great importance on the quality of the gate control circuitry, but also it is customary to place inductors in the d.c. leads connecting the bridge outputs to limit any transient circulating current should both bridges fire together.

A three-phase version of this arrangement, showing protective inductors, is given in *Figure 4.11.*

POWER CONTROL BY INTEGRAL CYCLING

All of the power control arrangements so far discussed in this book employ the control of a firing angle α in each half-cycle of supply. Another method involves switching on the thyristor(s) for a number of complete half-cycles (i.e. integral half-cycles) and then switching off for another number of complete half-cycles. Power is controlled by varying the on–off ratio of half-cycles. This type of control is also known as 'burst firing'.

Figure 4.12 shows a particular example, in which, out of a total control period of T cycles, the **on** period is N cycles.

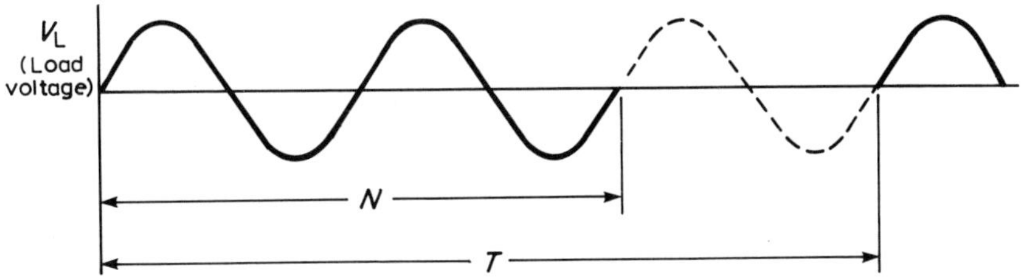

Figure 4.12 Integral cycling waveform

This type of control can be used for systems which are not sensitive to the intermittent nature of delivering power to the load. One example might be an electric furnace or kiln, in which the thermal time constant is long compared with the periodicity of the controlled supply power. It is not usual to control electric motors by integral cycling because of the very high starting currents which result from full cycle switching, nor is it desirable to control electric lamps in this way because of the sensitivity of the eye to the resulting flicker.

R.M.S. load voltage

Referring to *Figure 4.13*, the r.m.s. load voltage can be calculated by recognising that the total control waveform repeats every T cycles, and for N cycles the load voltage is the same as the supply.

Thus the base length is $2\pi T$ radians and the sinusoidal wave lasts for $2\pi N$ radians. Thus

$$V_{L(rms)} = \sqrt{\left(\frac{1}{2\pi T} \int_0^{2\pi N} V_m^2 \sin^2 \theta \, d\theta\right)}$$

which yields

$$V_{L\,rms} = \frac{V_m}{\sqrt{2}}\sqrt{\frac{N}{T}}$$

$$V_{L\,rms} = V_{rms}\sqrt{\frac{N}{T}}$$

where V_{rms} = r.m.s. supply voltage.

Load power (resistive load) For a resistive load such as a furnace, the load power is obtained from

$$P_L = \frac{V^2_{L(rms)}}{R_L}$$

Thus

$$P_L = \frac{V_m^2}{2R_L} \times \frac{N}{T}$$

$$= \frac{V_{rms}^2}{R_L} \times \frac{N}{T}$$

$$\therefore P_L = \frac{V_{rms}^2}{R_L} \times \frac{N}{T}$$

Power factor Using the general expression for power factor:

$$\text{p.f.} = \frac{\text{Actual input power}}{\text{Apparent input power}}$$

The active input power is equal to the r.m.s. load power, ignoring the small losses in the thyristors, and the apparent input power is the product of r.m.s. input voltage and r.m.s. input current. Thus

$$\text{Active input power} = \frac{V_{rms}^2}{R_L} \cdot \frac{N}{T}$$

and

$$\text{Apparent input power} = V_{rms} \times \frac{V_{rms}}{R_L}\sqrt{\frac{N}{T}}$$

since the input current must equal the load current. Then

$$\text{p.f.} = \frac{\dfrac{V_m^2}{2R_L} \cdot \dfrac{N}{T}}{\dfrac{V_m^2}{R_L}\sqrt{\dfrac{N}{T}}}$$

$$\therefore \text{p.f.} = \sqrt{\frac{N}{T}}$$

Worked example 4.3 A kiln controlled by integral cycling is worked at a duty cycle of 5:2 from a 240 V mains supply. The hot resistance of the kiln element is 12 Ω.

Determine the r.m.s. load voltage and the power factor presented to the supply.

$T = 7$ cycles and $N = 5$ cycles,

$$\therefore V_{L(rms)} = V_{rms}\sqrt{\frac{N}{T}}$$

$$= 240\sqrt{\frac{5}{7}}$$

$$= \underline{203\ \text{V}}$$

and

$$\text{p.f.} = \sqrt{\frac{N}{T}}$$

$$= \sqrt{\frac{5}{7}}$$

$$= \underline{0.85}$$

PROBLEMS FOR CHAPTER 4

(1) By reference to voltage and current waveforms describe the operation of a half-controlled thyristor bridge with an inductive–resistive load without a flywheel diode. Explain why a flywheel diode, if connected, produces no significant change in load waveforms for a bridge but modifies the conduction sequence.

(2) An electromagnet is controlled by a single thyristor from a 240 V r.m.s. supply. A flywheel diode is connected across the load. Assuming the load time constant is long and the coil resistance is 5 Ω, determine the mean load current for a firing angle of 0.5π radians.

Answer: 3.46 A

(3) Explain why it is advisable to use long duration trigger pulses for thyristors supplying inductive/resistive loads.

(4) A back-to-back thyristor pair supplies power to a load of time constant 10 ms. If the trigger pulses are of 1.5 ms duration determine the trigger angle α at which the circuit will begin to behave as a rectifier.

Answer: $\alpha \leqslant 0.25\pi$

(5) For the same firing angle, the measured mean d.c. load voltage for motor load is greater than for a static load. Explain with reference to a waveform sketch why this is so.

(6) Explain the mechanism of regenerative braking for a d.c. motor using two fully controlled thyristor bridges. Why is it usually necessary to include inductors between the bridges?

(7) A furnace controlled by integral cycling is to have a control range from zero to two-thirds full power. The resistance of the element in increases by 20% over this range, from a zero power value of 5 Ω, and at full power the resistance is 30% increased over its 'cold' value. If the control period T is 20 complete cycles, calculate the 'on' time (in cycles) per period for the two-thirds power condition. What power factor applies at this extreme?

Answer: Nearest value of $N = 12$
p.f. at $N = 12$ is 0.77)

5 Thyristor protection

The protection of thyristors means guarding against damage due to excessively high values of voltage or current, or to accidental triggering when the anode supply is suddenly applied, or to local overheating caused by exceedingly high rate of rise of anode current. This chapter deals with the principles of protection.

ABSOLUTE MAXIMUM RATINGS

Thyristors are sensitive devices. Even though they are capable of withstanding high values of voltage and passing large currents, if one of the maximum rating values is exceeded catastrophic failure in a very short space of time is likely to result. There are two sets of ratings – those concerning the anode–cathode path (or, in a triac, the terminal 1–terminal 2 path) and the other concerning the gate–cathode path.

Anode–cathode ratings

The main ratings are as follows:

V_D = maximum continuous off-state voltage (d.c.).
V_{DWM} = maximum crest working off-state voltage (a.c.).
V_R = maximum continuous reverse voltage (d.c.).
V_{RWM} = maximum crest working reverse voltage (a.c.).
$I_{T(RMS)}$ = maximum r.m.s. on-state current.
I_{TRM} = maximum repetitive peak on-state current.
$\frac{dV_D}{dt}$ = maximum rate of rise of forward anode voltage not to cause triggering.
$\frac{dI}{dt}$ = maximum permissible rate of rise of anode current.

Most of the above are self-explanatory, but it is worth noting that since many thyristor applications involve short conduction periods during alternate half-cycles of a.c. supply, the thyristor current waveform may be far from sinusoidal. For this reason, even though the r.m.s. thyristor current rating $I_{T(RMS)}$ may not be exceeded it is easily possible to exceed the repetitive peak current rating I_{TRM}. Typically, I_{TRM} lies between 5 and 10 times the value of $I_{T(RMS)}$.

The dV_D/dt rating comes about because of the anode–gate internal capacitance of the device. In effect, a very fast rate of rise of applied anode voltage, say during initial switch-on of a system, results in a gate voltage via the internal capacitance sufficient to turn the device on. Typically dV_D/dt ratings of 100–200 V/μs are quoted.

Lastly, the dI/dt rating results from the way the current carriers spread through the semiconductor regions. During device turn-on, conduction does not occur simultaneously over the whole region of the cathode, but concentrates first over a small area, near the gate, then spreads across the cathode. If the initial rate of rise of current is very large this means a small region is conducting at a very high current density at first, resulting in overheating and device failure. Figures of 20–200 A/μs are common.

Gate–cathode ratings

These are fairly straightforward, the main ones being as follows:

V_{RGM} = maximum reverse gate voltage.
I_{FGM} = maximum forward gate current.
P_{GM} = maximum peak gate power.
$P_{G(av)}$ = maximum average gate power.

It should be noted that, for triacs, the terms 'forward' and 'reverse' become meaningless because the device is designed to conduct under both conditions. The same general principles apply, however.

Thermal ratings

Because there is always a small p.d. across a conducting thyristor (typically 1–2 V) there will be internal heat generation and the semiconductor crystal will run hot. There is an upper temperature limit (typically 125°C) which is determined by the need to prevent a deterioration of the forward breakover voltage value, or by thermal instability within the crystal. The crystal is bonded to the metal body of the thyristor casing, which helps to reduce the crystal temperature, but usually it is necessary to bolt the thyristor to a large heat sink to dissipate the heat generated.

PROTECTION AGAINST VOLTAGE TRANSIENTS

Figure 5.1 shows some of the voltage phenomena which occur on a normal 50 Hz main supply, together with the switch-on conditions when, for example, a contactor is used to connect the main supply to the thyristor circuit and load.

During section A of the figure, due to contact bounce the thyristors experience some very rapid rises of anode voltage, which may exceed the dV_D/dt rating.

The regular spikes B add to the normal crest value of the supply and this affects the forward and reverse crest working voltage ratings.

The occasional surge C creates difficulties both of a dV_D/dt nature and a maximum forward (or reverse) voltage nature. It is good practice to use thyristors with a crest working voltage of at least twice the supply peak value to reduce the problems posed by the size of the surge C.

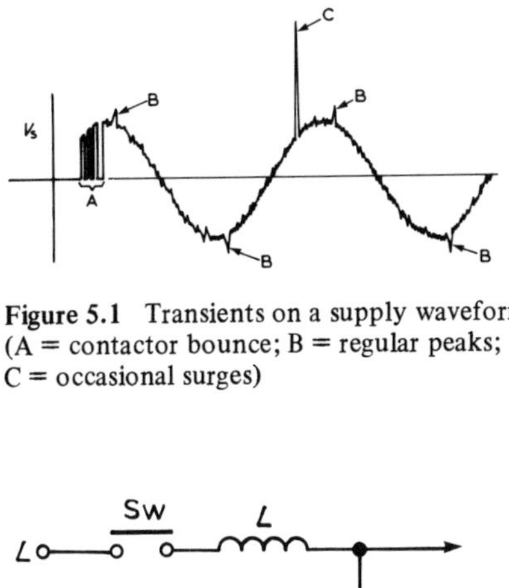

Figure 5.1 Transients on a supply waveform (A = contactor bounce; B = regular peaks; C = occasional surges)

Protection circuitry is recommended to slow down the rate of rise of anode voltage to acceptable values and at the same time remove or reduce the transient surges such as C in *Figure 5.1.*

Figure 5.2 shows a filtering arrangement commonly used in thyristor circuits in which the neutral is solidly connected but the mains isolation is in the line. L, R_1 and C form a damped series resonant circuit and R_2 provides a discharge path for C.

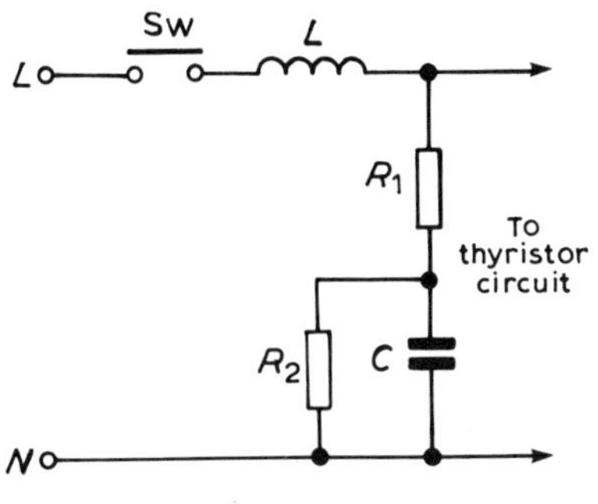

Figure 5.2 Transient filter circuit

A short duration surge such as that in *Figure 5.1* is very effectively filtered out by this arrangement, and indeed if R_2 is made zero practically nothing remains. However, when the contactor Sw is closed the sharp rise in voltage causes the input filter to *ring.* R_2 has to be of such a value as to limit the amplitude of this *ringing* (by damping) otherwise this may exceed the thyristor ratings. Thus the design of an input filter must be a compromise. It is made rather more difficult by the imprecise nature of the supply transients and the supply line's own inductance and resistance. However, if L is made large enough to swamp the line inductance some design procedures become possible (see Reference 1 at end of chapter).

A further problem occurs if the neutral is also switched (as in the case of a two-pole contactor connecting the supply to the thyristor circuit).

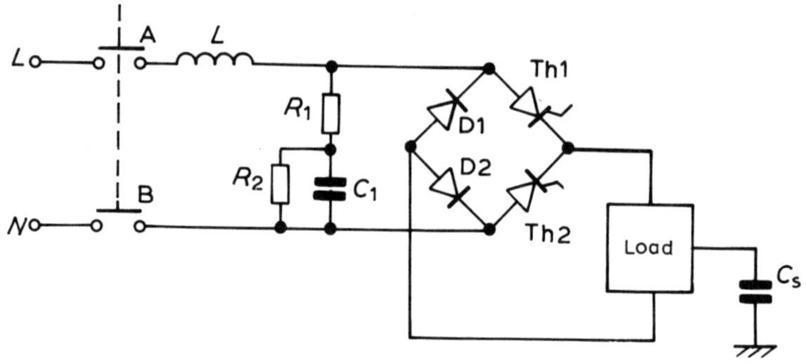

Figure 5.3 Line and neutral switching

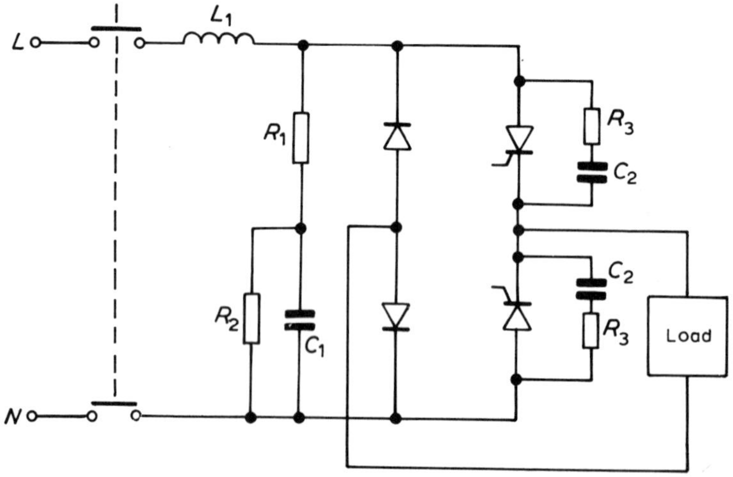

Figure 5.4 Full suppression circuit (L_1 = 250 μH, non-saturating; R_1 = 8.2 Ω, 3 W; R_2 = 33 kΩ, 3 W; R_3 = 10 Ω, 2 W; C_1 = 4 μF, 250 V a.c.; C_2 = 0.22 μF, 250 V a.c.)

Figure 5.3 shows such an arrangement, and the component C_s represents the stray capacitance to earth of the load.

Suppose, when the contactor is operated, contact A closes before contact B, and also suppose at this instant the line voltage is near the positive crest value.

Since $C_1 \gg C_s$ both supply lines are at the same a.c. potential ($\simeq + V_m$) at this instant and the diodes are reverse biased since their anodes are connected via C_s to earth. At this moment, therefore, both thyristors have practically the peak mains voltage suddenly applied to them in the forward direction in series with C_s to earth.

The dV/dt rate will be very high and either or both thyristors may trigger, supplying charging current to C_s. Of course, this current could not sustain the thyristors in conduction, but if, immediately this has happened, the neutral line contact B closes, thyristor Th1 will now conduct through the load and D2 to the neutral. If the load is a stationary motor, a damaging part cycle of conduction will flow and Th1 may well be destroyed. Fortunately the cure is quite simple. Each thyristor has a capacitor connected across it to slow down the dV/dt felt by the device. In practice, a series resistor/capacitor combination is used so that when the thyristor triggers under normal working conditions the discharge current from the capacitor through the thyristor is limited to a safe value.

A circuit showing full suppression for a single-phase supply feeding a thyristor half-controlled bridge is shown in *Figure 5.4*.

This arrangement is suitable for resistive or motor armature loads, but rather more involved circuitry is required for very highly inductive loads, or for three phase operation (see Reference 1).

FUSING

In common with all semiconductor junction devices, thyristors have a low heat capacity and they are therefore very susceptible to overloads. Two types of overload conditions exist, these being the short-circuit condition in which the current through the thyristor rapidly rises to a high value, and the less severe overload where the current exceeds the continuous rating of the device but not sufficiently to cause immediate breakdown. The first (short-circuit) condition requires very fast acting fuses for protection, capable of interrupting the circuit in less than 10 ms, since under short-circuit conditions the current will rise to a destructively high value during the course of a half-cycle of supply. The second condition is usually overcome by fast-acting overload contactors, or by electronic circuitry which detects the overload and then inhibits triggering of the thyristors. In this section the requirements of short-circuit protection will be considered.

Fusing characteristics

A modern fast-acting fuse is shown schematically in *Figure 5.5.* Within the ceramic body the fuse element (usually pure silver) is embedded in quartz sand or some other special quenching material. During normal operation there will be some heat generated, especially at the constrictions (necks) and this is conducted away via the sand and the large metal contact lugs.

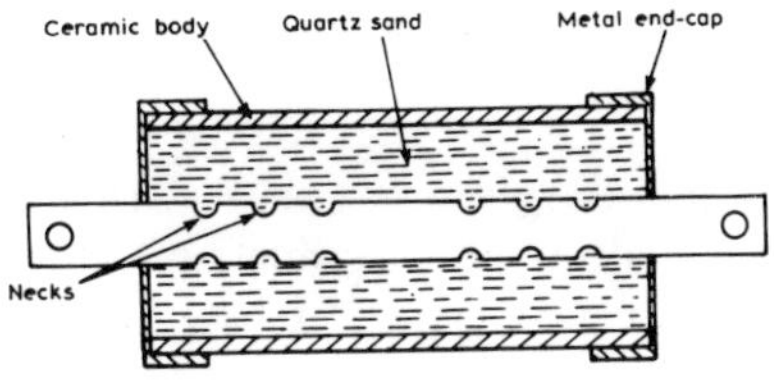

Figure 5.5 Construction of a fast-acting fuse

When a short-circuit occurs the temperature of the necks rises very rapidly and the fuse element begins to melt at these points. As soon as melting has occurred an arc will develop and will be sustained by the stored energy in the line inductance. The temperature at the arc points is very high, and as a result the quartz sand sinters, producing a non-conducting material which quenches the arc.

The total time for the fuse to operate is the sum of the *melting time* during which the short-circuit current is rapidly rising and the *arcing time* during which the current is falling as more and more sintering takes place.

Figure 5.6 shows an idealised fusing characteristic, superimposed on the supply waveforms of V and I.

In *Figure 5.6* the short-circuit occurs at time t_0 within the half-cycle; the normal load current at this time is I_0. The short-circuit current rises from I_0 towards a peak prospective value I_{PP}, on a curve determined by the impedance of the line. During the period t_0 to t_1 the fuse is melting and current is rapidly rising. At t_1 an arc is established and during the period t_1 to t_2 the arc is being strangled by the sintering sand. The period from t_0 to t_2 is the *clearing time* of the fuse, and the value of I at which melting is complete is $I_{fu(M)}$ (maximum fuse current let-through). Soon after the arc is established the voltage across the fuse reaches a high value (V_{ARC}), the size of which depends on the point in the supply waveform at which arcing commences.

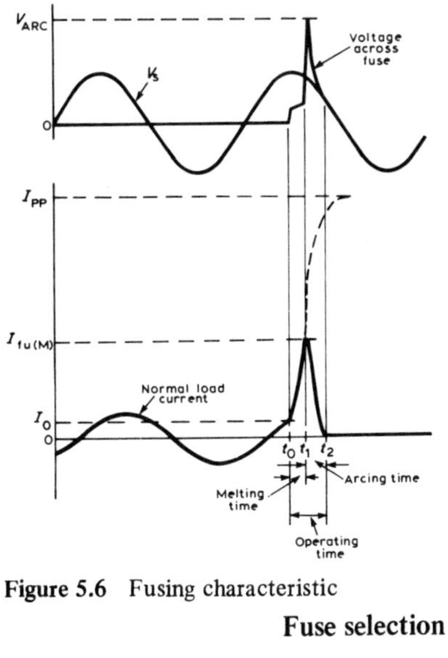

Figure 5.6 Fusing characteristic

Fuse selection

There are a number of factors to consider in choosing fuses for short-circuit protection of semiconductors; the main ones are given below.

1. *R.M.S. current rating.* The fuse selected must be capable of conducting the r.m.s. rated current for the thyristors, which is given in the thyristor data. (Alternatively, selection can be made on the basis of the r.m.s. current of the actual installation, but this will be within the thyristor ratings).
2. *I^2t value.* The I^2t value for a thyristor is a measure of the amount of electrical energy (which converts to heat in the semiconductor) which will not destroy the crystal. Manufacturers quote I^2t values for 10 ms since that time is a half-cycle of supply, and also in such a short time

the amount of heat loss is slight so that the same value holds for anything up to 10 ms. The fuse selected should have a lower I^2t value than that quoted for the thyristor it protects.

3. *Surge current rating (I_{TSM}).* The maximum surge current rating quoted by manufacturers is the peak value of a half-cycle of current which is the largest safely conducted by the thyristor. Since the fusing current characteristic (*Figure 5.6*) is roughly triangular, its peak value is rather short-lived compared with a sine wave. Thus $I_{fu(M)}$ can safely be higher than I_{TSM} and usually it is safe to work up to a value given by

$$I_{fu(M)} \simeq 1.4 I_{TSM}$$

4. *Peak fuse voltage rating.* If this is exceeded during a fault, arcing time will increase and the fuse will not clear quickly. Also, if the fuse has a peak voltage rating far in excess of the circuit voltage it will clear very quickly causing a sharp transient which may produce large voltage surges elsewhere in the circuit. These are potentially damaging to other semiconductors. It is good practice to select a fuse whose arc voltage lies between the peak applied voltage and the peak transient reverse voltage rating of the thyristor.

THERMAL PROTECTION – HEAT SINKS

The upper limit of junction temperature for thyristors is 125°C. Heat is conducted away, first, through the crystal and its encapsulation to the mounting base through a thermal resistance R_{jb} (junction–base thermal resistance). From there it is conducted into the heat sink through a thermal resistance R_{bs} (base–sink thermal resistance) and then it is conducted into the surrounding air through a thermal resistance R_{sa} (sink–ambient thermal resistance).

Figure 5.7 shows the various components in this process, with P_{TOT} representing the power dissipated by the device.

The values of R_{jb} and R_{bs} are determined by the construction of the device (size of silicon wafer, method of bonding to the case, size and type of mechanical fixing to heat sink) and are typically 0.6°C/W and 0.2°C/W, respectively.

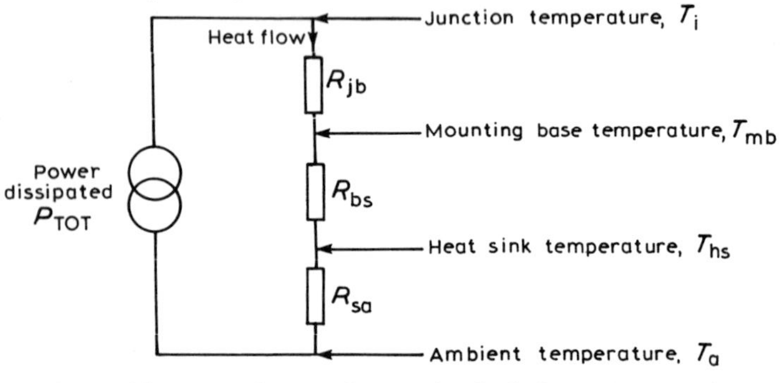

Figure 5.7 Components of thyristor thermal resistance

The problem is to design a heat sink which for a given maximum ambient temperature will keep the junction temperature below 125°C, or for a particular design determine how changes of conduction angle affect the permissible ambient temperature.

In *Figure 5.7* thermal resistance is analogous to electrical resistance, temperature is analogous to voltage and power is analogous to current; thus, for example,

$$R_{bs} = \frac{T_{mb} - T_{hs}}{P_{TOT}}$$

The following example will help to fix ideas on heat sink design.

Worked example 5.1 The maximum power dissipation for a thyristor is 60 W. Its thermal resistance values are

$$R_{jb} = 0.6°C/W$$
$$R_{bs} = 0.2°C/W$$

Determine the case temperature at maximum conditions and the maximum thermal resistance of a heat sink suitable for an environmental temperature of 35°C maximum.

Case temperature:

$$R_{jb} = \frac{T_j - T_{mb}}{P_{TOT}}$$

$$\therefore 0.6 = \frac{125 - T_{mb}}{60}$$

$$\therefore T_{mb} = 125 - 36 = \underline{89°C}$$

Heat sink thermal resistance:

$$R_{bs} + R_{sa} = \frac{T_{mb} - T_a}{P_{TOT}}$$

$$\therefore R_{sa} = \frac{T_{mb} - T_a}{P_{TOT}} - R_{bs}$$

$$= \frac{89 - 35}{60} - 0.2 = \underline{0.7°C/W}$$

Clearly, the next step is to consult the published curves for heat sinks of various finishes and geometries to finalise the design. Here, however, the situation becomes complicated because the thermal resistance of a specimen heat sink is not a constant. As its temperature above ambient rises, heat is lost more and more by radiation from its surface as well as by the process of air convection. Thus the thermal resistance falls as the temperature above ambient rises. Many manufacturers publish curves giving thermal resistance at a specified temperature above ambient, typically 40°C. *Figure 5.8* shows a typical set of such curves, relating to four types of sink made from extruded aluminium with blackened surfaces and mounted vertically in free air.

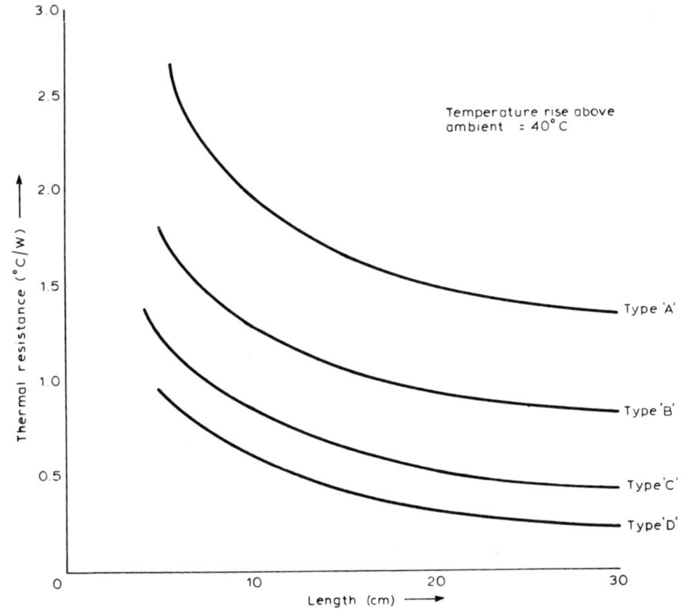

Figure 5.8 Heat sink characteristics

Reconsidering the above example in the light of *Figure 5.8* we have $T_{hs} - T_a = 40°C$ for the curves to be meaningful. But T_{hs} is found from

$$R_{bs} = \frac{T_{mb} - T_{hs}}{P_{TOT}}$$

i.e.

$$\begin{aligned} T_{hs} &= T_{mb} - P_{TOT} \times R_{bs} \\ &= 89 - 60 \times 0.2 \\ &= \underline{77°C} \end{aligned}$$

Thus the maximum permissible ambient temperature is

$$\begin{aligned} T_{a(max)} &= 77 - 40 \\ &= \underline{37°C} \end{aligned}$$

and the heat sink required is found from

$$\begin{aligned} R_{sa} &= \frac{T_{hs} - T_a}{P_{TOT}} \\ &= \frac{40}{60} \\ &= 0.67°C/W \end{aligned}$$

which corresponds to a length of 14 cm (approx) of type 'C' sink material.

Effect of conduction angle on heat sinks

The design outline given above relies on a knowledge of the actual dissipation P_{TOT} occurring in the thyristor, and is based usually on the worst-case assumption that the thyristor conducts for 180° (the complete half-cycle).

If, as is usually the case, the thyristor is required to conduct for less than 180°, the value of P_{TOT} falls which means, for a particular heat sink, the maximum permissible ambient temperature can rise. Alternatively a design may take as its starting point a known maximum conduction angle and deduce P_{TOT} for this before proceeding to determine the required heat sink.

Figure 5.9 shows a typical family of curves for a high power thyristor relating the total dissipation to the average current ($I_{av(\alpha)}$) with trigger angle α as a third variable.

Worked example 5.2

The thyristor for which *Figure 5.9* applies has the following characteristics:

$$\begin{aligned} I_{av(\alpha)max} &= 90 \text{ A} \\ T_{jmax} &= 125°C \\ R_{jb} &= 0.3°C/W \\ R_{bs} &= 0.1°C/W \end{aligned}$$

Using the heat sink data of *Figure 5.8* design a sink for use at 180° conduction angle ($\alpha = 0°$) for a 90 A average thyristor current and determine the maximum ambient temperature allowed. Assuming the

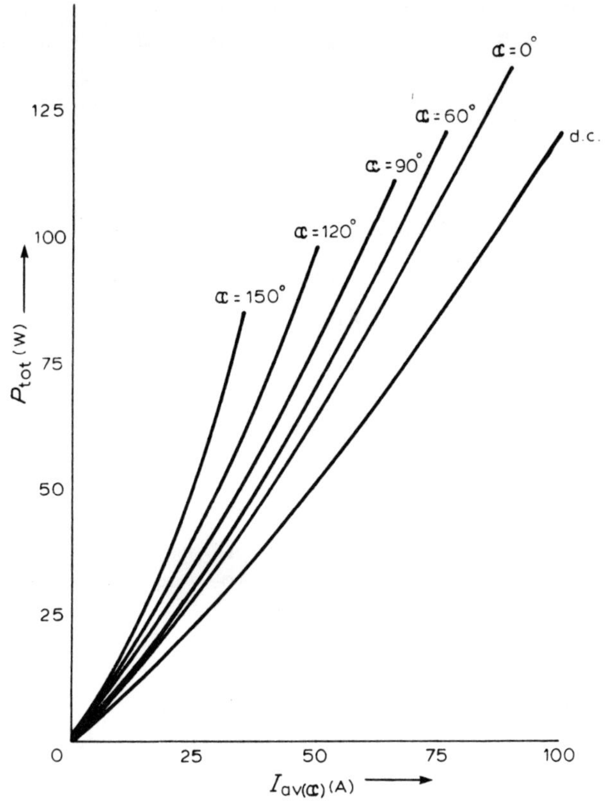

Figure 5.9 Effect of trigger angle on dissipation

load is a constant resistance, determine the maximum ambient temperature for this design if the conduction angle is reduced to 90° ($\alpha = 90°$).

From *Figure 5.9*, at 90 A and $\alpha = 0°$, the dissipation is

$$P_{TOT} = 137 \text{ W}$$

Thus

$$R_{sa} = \frac{40}{137} \simeq 0.3°\text{C/W}$$

A 21 cm length of type 'D' will suffice for this.

From

$$R_{jb} + R_{bs} + R_{sa} = \frac{T_s - T_a}{P_{TOT}} \quad (1)$$

then

$$0.3 + 0.1 + 0.3 = \frac{125 - T_a}{137}$$

$$\therefore T_a = 125 - 96$$

$$= \underline{29°\text{C}}$$

If $\alpha = 90°$, then the average current will be half of its value at $\alpha = 0°$ since the load resistance is unchanged.

$$\therefore I_{av(90)} = 45 \text{ A}$$

and

$$P_{TOT(90)} = 70 \text{ W}$$

from *Figure 5.9.* Thus, using equation (1) again with the new value of P_{TOT},

$$0.7 = \frac{125 - T_a}{70}$$

$$\therefore T_a = 125 - 49$$
$$= \underline{76°C}$$

Note: A heat sink of the kind in this example would have many large fins and would be of thick aluminium. Probably there would also be forced cooling.

PROBLEMS FOR CHAPTER 5

(1) The maximum dissipation for a thyristor is 5 W when its mounting base temperature is 105°C. If the value of R_{bs} is 1.3°C/W determine:

(a) The value of R_{jb}.
(b) The heat sink temperature.
(c) The value of the thermal resistance of the heat sink if it is operating at 40°C above ambient.
(d) The maximum ambient temperature.

Answer: (a) 4°C/W; (b) 98.5°C
(c) 8°C/W; (d) 58.5°C

(2) The thermal resistances of two heat sinks of identical geometries and materials are identical at low temperatures but diverge greatly at higher temperatures. One has a matt black surface finish and the other has a bright metal finish. Explain why the divergence occurs and identify the sink with the lowest thermal resistance.

(3) Two thyristors are mounted close together on a single heat sink. The data for each is as follows:

	Thyristor 1	*Thyristor 2*
R_{jb}	0.6	0.5
R_{bs}	0.3	0.2
P_{TOT}	20 W	30 W

Determine which thyristor has the higher junction temperature and deduce from this the maximum junction temperature of the other. Using the curves of *Figure 5.8*, determine a suitable heat sink and calculate the maximum permissible ambient temperature (assume that the heat sink behaves as though one device equivalent to the sum of the dissipations is bolted to it).

Answer: (a) Thyristor 2 has the higher junction temperature
(b) $T_{j1(max)} = 122°C$
(c) 6.5 cm type 'B' or 16 cm type 'A'
(d) $T_{amb(max)} = 64°C$

(4) A thyristor system is designed to operate at a maximum conduction angle of 120° ($\alpha = 60°$). Each thyristor contributes a mean current of 50 A at this value of α, and the curves of *Figure 5.9* apply to the thyristors used. Determine a suitable heat sink and the maximum ambient temperature using the curves of *Figure 5.8*, given the following:

Maximum mounting-base temperature = 95°C

$$R_{bs} = 0.1°C/W$$

Answer: (a) 11 cm type 'D' or 18.5 cm type 'C' } $R_{sa} = 0.55°C/W$
(b) $T_{amb(max)} = 47.75°C$

(5) Determine, for the system in Problem (4), the maximum ambient temperature if the conduction angle is reduced to 60° ($\alpha = 120°$), assuming a constant resistive load.

Answer: $I_{av(120)} = 16.5$ A
$P_{TOT} \simeq 25$ W
$T_{amb(max)} = 101°C$

REFERENCE

1. *Power Engineering Using Thyristors*, Mullard Technical Publications.